Planets and Planetary Systems

Planets and Planetary Systems

Stephen Eales

School of Physics and Astronomy, Cardiff University UK

WILEY-BLACKWELL

A John Wiley & Sons, Ltd., Publication

This edition first published 2009

Wiley-Blackwell is an imprint of John Wiley & Sons, formed by the merger of Wiley's global Scientific, Technical and Medical business with Blackwell Publishing.

Registered office: John Wiley & Sons Ltd, The Atrium, Southern Gate, Chichester, West Sussex, PO19 8SQ, UK

Other Editorial offices: 9600 Garsington Road, Oxford, OX4 2DQ, UK
111 River Street, Hoboken, NJ 07030-5774, USA

For details of our global editorial offices, for customer services and for information about how to apply for permission to reuse the copyright material in this book please see our website at www.wiley.com/wiley-blackwell

Library of Congress Cataloging-in-Publication Data:

Eales, Stephen.
Planets and planetary systems / Stephen Eales.
p. cm.
Includes bibliographical references and index.
ISBN 978-0-470-01692-3 (cloth) – ISBN 978-0-470-01693-0 (pbk.) 1.
Planetary theory. 2. Planets. 3. Solar system. I. Title.
QB361.E35 2009
523.4–dc22

2008055970

ISBN: 9780470016923 (HB) and 9780470016930 (PB)

A catalogue record for this book is available from the British Library.

Typeset in 10.5/13 Minion by Laserwords (Private) Ltd, Chennai, India
First impression 2009

Contents

Preface

The ideal planetary scientist would have knowledge of astronomy, physics, chemistry, geology, meteorology, oceanography and, because both the atmosphere and the surface of our planet have clearly been extensively modified by living creatures, biology. Although I have given a course on planets and planetary systems for the last decade, I can only really claim to be an expert in two of these areas, but the liberating thing about writing a book on such a huge rambling interdisciplinary subject is that nobody else has the perfect credentials for writing one either. As with many writers of textbooks, I decided to write this book, not because I wanted to share my wisdom with the world, but because I never found a textbook that was perfect for my course. The available books were either too basic or were graduate-level tomes much too big (and expensive) for an undergraduate course, exacerbated by the fact that planetary science is such a dynamic area of research that any textbook gets out of date very quickly. Although I am an astrophysicist rather than an oceanographer or a geologist, I have tried to write a general introduction to planets and planetary systems that uses insights from *all* the disciplines involved in the study of these objects. The book should be suitable for any student studying planets or planetary systems as part of an undergraduate science degree, and I have also provided a less mathematical route through the book for any student that does not have a basic knowledge of calculus (an elementary knowledge of differentiation and integration).

In such a rapidly changing field, I have tried to make the book as up to the minute as possible by incorporating results from the most recent planetary space missions, such as the Cassini mission to Saturn and the many recent missions to Mars. I have also listed recent scientific papers as further reading at the end of some of the chapters; since these are mostly taken from the journals *Science* and *Nature*, which, at least in intention, are journals for the non-specialist reader, they should be comprehensible to any undergraduate. Nevertheless, any book in such a rapidly changing subject gets out of date very fast. If you are still interested in planetary science after finishing this book, there are a number of ways you can learn about new discoveries in the field. The best place to look, of course, is the internet. Every space mission has a web site, and once you know the name of the space mission, it is easy to find the web site using a standard search engine (Appendix 1 contains a list of past space missions and a provisional list of upcoming space missions). There are also two valuable databases of scientific papers on the internet. The astrophysics preprint database (http://xxx.lanl.gov/archive/astro-ph) is an archive of astronomy papers written since April 1992, although unfortunately planetary scientists have been slower than other groups of astronomers in using the archive. The NASA Astronomy Data System (http://adsabs.harvard.edu/abstract_service.html) is an archive of all astronomy papers that have *ever* been published. This is now an essential resource for any astronomer; it is possible, for example, to use it to find all the papers that have ever been published on any subject in which you are interested and to read papers that were written decades ago by some of the giants in the field, for example Oort's original paper on the Oort Cloud.

I hope you enjoy the book. Please e-mail me any comments or suggestions for improvements for future editions.

Stephen Eales
sae@astro.cf.ac.uk

Physical Constants

Symbol	Value in SI units	Meaning
c	2.9979×10^{8} m s^{-1}	Speed of light
G	6.670×10^{-11} m^{3} kg^{-1}s^{-2}	Gravitational constant
h	6.626×10^{-34} J s	Planck's constant
k	1.381×10^{-23} J K^{-1}	Boltzmann's constant
m_{e}	9.109×10^{-31} kg	Mass of electron
m_{p}	1.673×10^{-27} kg	Mass of proton
m_{amu}	1.661×10^{-27} kg	Atomic mass unit
N_{A}	6.022×10^{23} mol^{-1}	Avogadro's number
σ	5.667×10^{-8} W m^{-2} K^{-4}	Stefan–Boltzmann constant

Astronomical Constants

Symbol	Value in SI units	Meaning
AU	1.496×10^{11}m	Earth-Sun distance
Pc	3.086×10^{16} m	Parsec – astronomical unit of distance
$M_{\odot}$	1.989×10^{30} kg	Solar mass – astronomical unit of mass
$L_{\odot}$	3.827×10^{26} W	Solar luminosity – astronomical unit of luminosity

1 Our planetary system

World is crazier and more of it than we think,
Incorrigibly plural. I peel and portion
A tangerine and spit the pips and feel
The drunkenness of things being various
Louis MacNeice

1.1 Diversity in the Solar System

We are living during one of the great periods of human exploration. During the last few decades, and continuing today, the human species is exploring its planetary system for the first time – an exciting period that will only happen once. A generation ago, Mercury, Neptune and Uranus were just points of light; the Edgeworth–Kuiper belt (EK belt) had not yet been discovered; only 13 moons of Jupiter were known (the current count is 63); and nobody had any idea what lay below the clouds of Venus – and this is just to give a few examples. As I write this book, the exploration of our planetary system is entering a new intense phase. Four space missions are currently exploring Mars, and over the next two decades space missions will be visiting Mars virtually every year, in preparation for the first human landing, which may be sometime around the year 2030. As for the rest of the Solar System, the Cassini spacecraft is currently cruising among the moons of Saturn and spacecraft are on their way to Mercury (Messenger), the asteroid belt (Dawn) and Pluto (New Horizons) (see Appendix 1). One of the big discoveries from this epoch of exploration is that all the planets are very different. When the planets were just points of light, it was possible to imagine that they might actually be quite similar, but we now know that each planet is immediately recognizable and very different from all the others. One of my goals in this chapter is to consider some of the reasons for this amazing diversity.

Planets and Planetary Systems Stephen Eales

In the same period that we have learned so much about our own planetary system, we have begun to learn about other planetary systems. Only a decade ago, the only planetary system we knew about was our own. There are now almost 200 planets that have been discovered around other stars. All of these are giant planets, but both the European Space Agency (ESA) and the National Aeronautics and Space Administration (NASA) are designing space missions that will be able to observe planets as small as the Earth. It is already clear that these other planetary systems are often very different from our own (Chapter 2), and therefore planetary systems as well as planets are very diverse. These other planetary systems are interesting in themselves, but they are also important because they allow us to see our own planetary system in context. Is the solar system an unusual or a run-of-the-mill planetary system? The answer to this question is important because of one of the most interesting facts about the solar system – the fact that it harbours life. If the solar system is a typical planetary system, one might expect that life would be a fairly common phenomenon in the universe. As I will describe later in this book, the future NASA and ESA missions will be able to search for life on any of the planets they discover, by looking for atmospheric gases that are the product of living organisms.

Let us start with a quick tour of our own planetary system. The closest planet to the Sun is Mercury. Until recently, almost everything we knew about this planet came from the Mariner 10 spacecraft, which flew past Mercury in 1972, photographing 40 % of its surface. In January 2008 the American Messenger spacecraft took the first new images of the planet for over three decades when it flew past Mercury on a complicated voyage – it will fly past Mercury three times, Venus twice and the Earth once – which will ultimately put it in orbit around the planet in 2011 (Figure 1.1). Both the Mariner 10 and Messenger images show a rocky surface covered in craters resembling the surface of the Moon. The instruments on Mariner 10 revealed that the planet has virtually no atmosphere and that it has a magnetic field, and another result of this mission was the first measurements of the planet's mass and density. Such basic measurements may not sound an impressive scientific achievement, but it is impossible to measure the mass and density of a moonless planet without sending a spacecraft there, because a planet's orbit around the Sun is independent of its mass. For a circular orbit, the gravitational force between the planet and the Sun must equal the centripetal force:

$$\frac{GM_{\mathrm{S}}M_{\mathrm{p}}}{d^2} = \frac{M_{\mathrm{p}}v^2}{d} \tag{1.1}$$

In this equation M_{p} and M_{S} are the masses of the planet and the Sun, d is the distance between them, and v is the speed of the planet. The planet's mass appears on both sides of the equation and so cancels out, which means the planet's speed is independent of its mass. The only way to measure the mass of a planet without a moon is to measure the trajectory of a spacecraft as it passes close to the planet.

(a)

(b)

(c)

Figure 1.1 The eight planets in our planetary system. (a) Mercury (Messenger); Venus (Pioneer Venus Orbiter); Earth (Apollo 8). (b) Mars (Viking Orbiter); Jupiter (Voyager 2); Saturn (Voyager 2). (c) Uranus (Voyager 2); Neptune (Voyager 2) (courtesy: NASA). A colour reproduction of this figure can be seen in the colour section, located towards the centre of the book.

Once one has measured the mass of a planet, one can calculate its density. Mercury has the second highest density of the planets in the solar system (Table 1.1), and this fact, together with the existence of a magnetic field, have led scientists to conclude that it has an iron core.

Venus, the second planet from the Sun, used to be a favourite location for science fiction writers because the clouds hiding its surface made it possible to imagine any kind of life there (dinosaurs roaring through primeval swamps was one idea). As the result of probes, mostly Russian, that have descended through the clouds,

Table 1.1 Some properties of the planets.

(1) Name	(2) Distance from Sun (AU)	(3) Number of moons	(4) Mass ($\times 10^{24}$ kg)	(5) Density (kg m^{-3})	(6) Observed Temperature (K)	(7) Temperature Predicted from Equation 1.5 (K)
Mercury	0.39	0	0.33	5.4	100–725*	451
Venus	0.72	0	4.87	5.2	733	260
Earth	1.00	1	5.97	5.5	288	255
Mars	1.52	2	0.64	3.9	215	222
Jupiter	5.20	63	1898.6	1.3	124	104
Saturn	9.54	56	568.5	0.69	95	79
Uranus	19.2	27	86.8	1.32	59	58
Neptune	30.1	13	102.4	1.64	59	55

*The large range of temperatures for Mercury is the result of its slow rotation, which produces very low temperatures at night despite the planet's small distance from the Sun.

we now know that humans would be immediately killed in four different ways: asphyxiated by the lack of oxygen; broiled by the high temperature (about 730 K on the surface); crushed by the pressure (700 times the pressure of the atmosphere on the Earth); and finally dissolved by the rain of sulfuric acid that drizzles down from the Venusian sky. In the 1990s, the Magellan spacecraft mapped the planet by bouncing radio waves, which pass through the clouds, off the surface. The Magellan maps revealed many geological structures unlike any on Earth (Figure 1.2).

The third rock from the Sun is, of course, the Earth. Similar in mass and diameter to Venus, it is different in most other respects. The Earth's atmosphere is mostly composed of oxygen and nitrogen, whereas the atmosphere of Venus is almost completely composed of carbon dioxide. The oxygen in the Earth's atmosphere is the consequence of life, the product of photosynthesis in plants. If life suddenly vanished from the Earth, oxygen, which is a highly reactive element, would gradually disappear from the atmosphere by combining with other atmospheric gases and rocks. The existence of life makes the Earth unique among the planets (at least, as far as we know at the moment). It is also unique because it is the only planet with oceans and also the only one with a system of mobile tectonic plates. Life and the presence of liquid water are almost certainly connected. The connection between life and a system of tectonic plates is not so obvious, but it is possible that an active geological surface is part of the reason why the Earth's temperature has remained surprisingly constant for the last 4.5 billion years (Chapter 9).

Mars, the next planet, has always been a popular place to look for extraterrestrial life. Early in the last century, the astronomer Percival Lowell became convinced

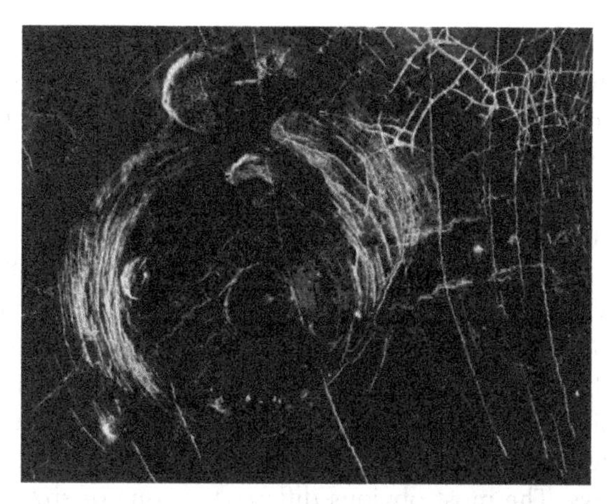

Figure 1.2 One of the strange geological features on the surface of Venus revealed by the radar on the Magellan spacecraft. The cause of these 'pancake domes' is not certain, but one possibility is that liquid rock (magma) under the surface pushed the surface outwards, and then sank back into the planet's interior, causing the surface to collapse (courtesy: NASA).

that he could see canals on the planet, which he thought might be an attempt by a dying civilization to transport water from the planet's polar caps. We now know, because the space missions to the planet have not seen them, that the canals were an optical illusion brought on by wishful thinking. However, conditions on the planet were once probably suitable for life. The same space missions that disproved Lowell's canals have discovered many features that look like dried-up riverbeds (Figure 1.3) and gouges in the surface that look as if they have been carved by flash floods. The European mission Mars Express has shown there is a large reservoir of ice in the polar caps and discovered a possible dust-covered frozen sea close to

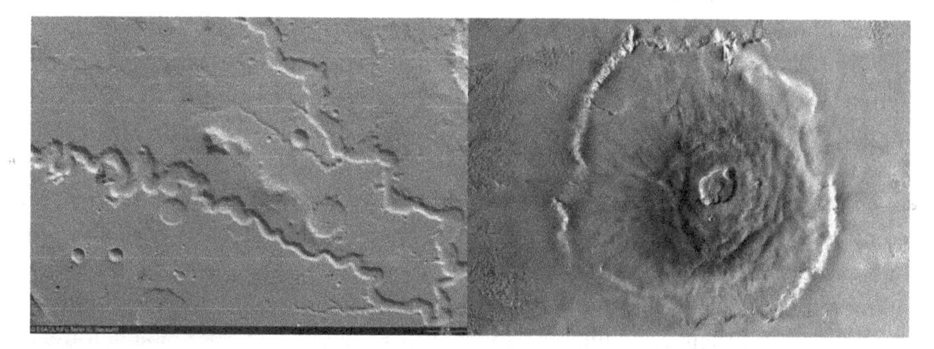

Figure 1.3 Two images of Mars: on the left, a possible dried-up riverbed (Mars Express, courtesy ESA); on the right, the largest volcano in the solar system, Olympus Mons (Viking Orbiter, courtesy: NASA).

the equator (Chapter 3). Instruments on the spacecraft have also revealed minerals on the surface that could only have been formed if Mars was once a wet planet. Mars currently only has a very tenuous atmosphere, composed mostly of carbon dioxide. However, the mass of Mars is only 11 % that of the Earth, and it is possible that Mars once had a much denser atmosphere and gradually lost it because of its relatively weak gravitational field. Another noteworthy thing about Mars is the size of its volcanoes. The largest volcano in the solar system (Olympus Mons) is on Mars (Figure 1.3). This measures 25 km from base to peak and has a diameter at its base of ~600 km, compared with the relatively puny Mauna Loa, the largest volcano on Earth, which has a height of 9 km and a base diameter of ~100 km.

We will consider the next four planets as a group: Jupiter, Saturn, Uranus and Neptune, in order from the Sun. They are different from the inner planets in several fundamental ways. The most obvious difference is one of size: the outer planets dwarf the inner planets. The largest, Jupiter, has a mass 300 times that of the Earth and even the smallest, Uranus, has a mass 15 times the Earth's. A second difference is implicit in the name that is often used for the outer planets: gas giants. Whereas the inner planets are essentially balls of rock surrounded by a very thin layer of gas, the outer planets are mostly atmosphere, and it is not even yet clear whether the outer planets contain any rocky core at all (Chapter 4).

A less obvious difference is in composition. The two principal methods that have been used to determine the overall chemical composition of the solar system are spectroscopy of the Sun and chemical analysis of primitive meteorites called *carbonaceous chondrites*, whose composition probably reflects that of the original solar nebula (Chapter 8). Both methods have advantages and disadvantages. An advantage of the latter is that it is possible to measure the abundances of the elements in a meteorite with great precision, but the disadvantage is that some of the volatile elements are probably missing. The advantage of the former is that the composition of the solar photosphere must be very similar to that of the solar system as a whole, but the disadvantage is that the abundance ratios that can be obtained from spectral lines are much less accurate than with the other method. Table 1.2 shows the abundances of the 10 most common elements in the solar system. The solar system is dominated by only two elements, hydrogen and helium, which contain about 98 % of the mass of all the elements combined. The Earth and the other inner planets are mostly made out of the elements that form the remaining 2 %: silicon, oxygen, magnesium and so on. The atmospheres of the outer planets, though, are composed mostly of the dominant two elements. They also contain small amounts of molecules such as hydrogen sulfide (H_2S), water (H_2O), methane (CH_4) and ammonia (NH_3), and it is these molecules that are responsible for the clouds on the planets and their very different appearances (Figure 1.1). The blue colours of Uranus and Neptune, for example, are caused by methane, which strongly absorbs

Table 1.2 The 10 most abundant elements in the solar system.

Element	Atomic number	Number of atoms of element relative to hydrogen
Hydrogen	1	1
Helium	2	0.085
Oxygen	8	4.6×10^{-4}
Carbon	6	2.5×10^{-4}
Neon	10	6.9×10^{-4}
Nitrogen	7	6.0×10^{-5}
Magnesium	12	3.4×10^{-5}
Silicon	14	3.2×10^{-5}
Iron	26	2.8×10^{-5}
Aluminium	13	2.3×10^{-6}

red light; the light we see, which is simply reflected sunlight, is thus missing the red end of the spectrum.

A final difference between the inner and outer planets is in the objects that surround them. Mars has two tiny moons, Venus and Mercury do not have moons at all, and the Earth has the only large moon in the inner solar system. All the outer planets have large numbers of moons, and they also all have rings, from the spectacular rings of Saturn, which are visible with even a small telescope, to the rings of Neptune that were only discovered by Voyager 2.

The moons in the solar system, like the planets, at first sight exhibit a bewildering range of properties. The six largest moons in the solar system are shown in Figure 1.4. Their images are very different and they are all immediately recognizable. The one with the large dark areas, which early astronomers thought were oceans, is of course our moon. Titan, the largest moon of Saturn, is the one covered in a haze and is the only moon with a substantial atmosphere. The lurid colours of Io, the closest moon to Jupiter, make it look remarkably like a pizza. Europa, the second moon out from Jupiter, has a smooth surface that is covered in fine lines; and the other two of Jupiter's giant moons, Ganymede and Callisto, if not so spectacular, also look completely different from all the others.

Apart from the eight planets, the solar system contains tens of thousands of smaller objects. Most of these orbit the Sun in two 'belts'. The asteroid belt was discovered in the nineteenth century and consists of at least 10 000 objects that orbit the Sun between the orbits of Mars and Jupiter. The largest of these is Ceres, which has a diameter of about 900 km; the smallest that have so far been discovered have a diameter of only a few kilometres. Figure 1.5 shows an image of the asteroid Ida, taken by the Galileo spacecraft on its way to Jupiter. The image shows that Ida,

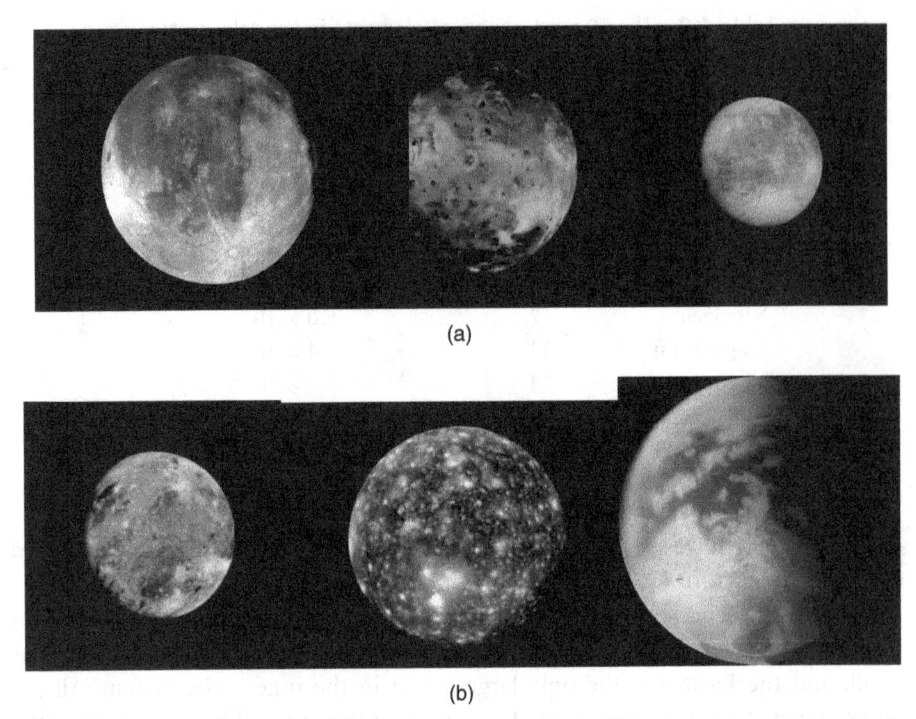

Figure 1.4 The six largest moons in the solar system (not to scale). (a) the Moon (Clementine); Io (Voyager 1); Europa (Voyager 1). (b) Ganymede (Voyager 1); Callisto (Voyager 1); Titan (Cassini) (courtesy: NASA and ESA). A colour reproduction of this figure can be seen in the colour section, located towards the centre of the book.

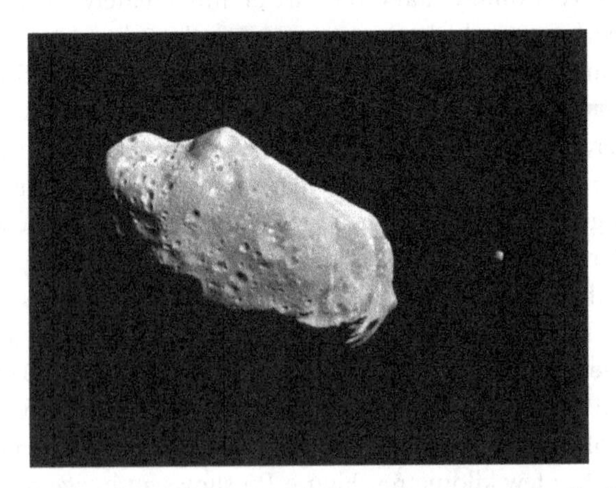

Figure 1.5 The asteroid Ida and its moon Dactyl, an image taken by the Galileo spacecraft (courtesy: NASA).

which is 56 km long, has a tiny moon, Dactyl, only about 1.5 km in size. Despite the large number of objects, the total mass in the asteroid belt is not very large, only $\sim 5 \times 10^{-4}$ of the mass of the Earth.

The second of the belts, the EK belt, was only discovered in 1992. This consists of objects that orbit the Sun outside the orbit of Neptune. There are currently about 1000 of these known, although because only a small part of the sky has been searched to the necessary sensitivity (these objects are small and a long way from the Sun and so are very faint), astronomers have estimated that there may be as many as 100 000 of them. The second largest of the objects in the EK belt is a very well known object, about which I will write more below.

1.2 General trends in the properties of the planets

Let us now consider some of the reasons for the rich diversity that we see within our planetary system. We will first consider the temperatures of the planets. Column 6 in Table 1.1 lists their approximate average temperatures.

Let us assume for the moment that the only heating source for each planet is the Sun. The planet's temperature will then reflect the balance between the energy it absorbs from the Sun and the energy it radiates. The power carried by the sunlight is $L_\odot/4\pi D^2)$ W m^{-2}, in which $L_\odot$ is the luminosity of the Sun and D is the planet's distance from the Sun. The cross-sectional area of the planet is πR_p^2, R_p being the radius of the planet, and so the power absorbed by the planet is

$$P_{abs} = \frac{L_\odot R_p^2(1-A)}{4D^2} \tag{1.2}$$

A is the albedo of the planet, which varies between 0 and 1, and is a measure of the fraction of the sunlight that is reflected back into space; the reflected light, of course, does not heat the planet. Values of the albedo for objects in the solar system range from 0.04 for a hemisphere of Iapetus, one of the moons of Saturn, which is as dark as lampblack, to 0.67 for Europa, a moon of Jupiter, which is covered in ice. The power a planet radiates from a square metre of its surface is $\varepsilon\sigma T^4$, which is just the Stefan–Boltzmann law for a black body multiplied by the planet's emissivity, ε. In the infrared waveband, in which the planets radiate most of their energy (see below), the value of ε is about 0.9. The total power radiated by the planet is thus

$$P_{rad} = 4\pi R_p^2 \varepsilon\sigma T^4 \tag{1.3}$$

In equilibrium, the power radiated by the planet equals the power absorbed from the Sun, and so

$$\frac{L_\odot R_p^2(1-A)}{4D^2} = 4\pi R_p^2 \varepsilon\sigma T^4 \tag{1.4}$$

If we rearrange Equation 1.4, we get the equation for the equilibrium temperature of a planet:

$$T = \left(\frac{L_{\odot}(1 - A)}{16\pi\varepsilon\sigma D^2} \right)^{\frac{1}{4}} \tag{1.5}$$

If I use the measured albedos of the planets in this equation, I predict the temperatures of the planets that are given in column 7 of Table 1.1. A moment's comparison between these and the observed temperatures shows that this fairly simple piece of physics gives a surprisingly good explanation of the planets' temperatures. The obvious exception is Venus, which is much hotter than predicted, although the Earth, Jupiter and Saturn are also a little hotter than predicted. The explanation of the discrepancy for Jupiter and Saturn is that these planets must also have an internal energy source, either the original heat which was stored in the planet when it was formed (Chapter 8), which is slowly leaking out, or the gradual conversion of gravitational potential energy into heat as denser material gradually settles towards the centre of the planet. As we will now see, the explanation of the discrepancies for Venus and the Earth is one of those surprising places where astronomy suddenly becomes quite relevant to human affairs.

We can determine the waveband in which the planets emit most of their radiation by using Wien's displacement law, which gives a relationship between the temperature of an object and the wavelength at which the luminosity of the object is at a maximum:

$$\lambda_{max} = \frac{0.029}{T} \tag{1.6}$$

in which the wavelength, λ_{max}, is measured in metres and the temperature, T, in Kelvin. The temperature of the Sun's photosphere is about 6000 K, and this law shows the wavelength at which the Sun's radiation is at a maximum is 0.48 μm, which, as one might expect, is in the optical waveband. The planets are much cooler than the Sun and application of Wien's law shows that they emit most of their radiation in the infrared waveband; for example λ_{max} for the Earth is 10 μm. The explanation of Venus' high temperature is its dense atmosphere of carbon dioxide. Its surface is heated by sunlight (actually mostly by the Sun's ultraviolet light because the optical light is blocked by the clouds). The surface emits infrared radiation, but this cooling radiation cannot escape through the atmosphere because carbon dioxide absorbs infrared radiation – and thus the surface heats up. For the reason that glass is also transparent to optical radiation but opaque to infrared radiation this phenomenon is called the '*greenhouse effect*'. The small amounts of carbon dioxide, water vapour and methane in its atmosphere also keep the Earth warmer than it would otherwise be, and a glance at Table 1.1 shows that this is a very good thing, because without these greenhouse gases the average temperature of the Earth would be well below the freezing point of water. The reason why this

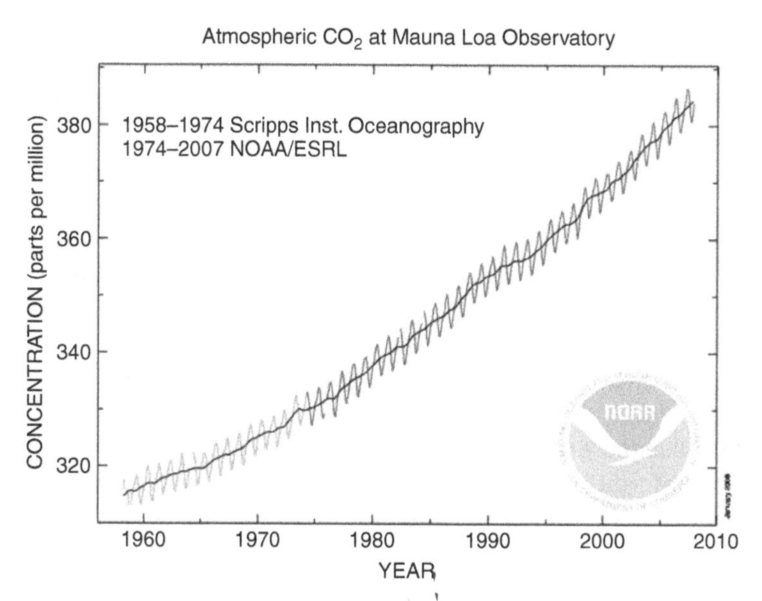

Figure 1.6 The concentration of carbon dioxide in the Earth's atmosphere in parts per million measured at Mauna Loa Observatory. The oscillation is due to the growth of plants during the summer removing carbon dioxide from the atmosphere, which is then returned by the decay of plants in the winter. Apart from this oscillation, the long-term trend is clearly upwards, almost certainly due to the burning of fossil fuels and deforestation (courtesy: Dr Pieter Tans, NOAA/ESRL).

bit of astronomy has more than an abstract interest, of course, is that in the future the greenhouse effect may well become a very bad thing, because of the increasing amount of carbon dioxide in the Earth's atmosphere produced from cars, factories and aeroplanes (Figure 1.6).

The other obvious trends in Table 1.1 are that the inner planets are denser and smaller than the outer planets. The difference in density is undoubtedly connected to the difference in composition, but this just alters the question to why the composition of the two sets of planets should be so different. The answer, as I will describe in detail in Chapter 8, is probably again the heating effect of the Sun. The planets formed out of a disc of gas, which was hotter at its centre because of the newly formed Sun. As the gas cooled, different chemical compounds began to freeze, and tiny solid particles began to appear within the gas, which eventually stuck together (the details of how they did this are still unclear) to form the planets. In the inner part of the disc, only compounds with high melting points froze, so it is not surprising that the inner planets are made out of compounds with high melting points. The difference in the masses of the two sets of planets is harder to explain, especially because the other planetary systems that have so far been discovered contain giant planets that are very close to their stars (Chapter 2). Nevertheless, with some elaboration, the standard model for the formation of a planetary system can explain both planetary systems that have giant planets very close to their stars and planetary systems like ours with the giant planets much further from their stars, although many of the details of this explanation are still unclear (Chapter 8).

It is possible to use a neat physical argument to explain one other interesting difference between the inner planets: the ages of their surfaces. All the inner planets and also the Moon show signs on their surfaces of some geological activity (Figures 1.2 and 1.3), although unless one can actually see a volcano erupting, as on the Earth, it is not usually obvious whether this geological activity is occurring today or whether it happened billions of years ago. I will show in Chapter 3 how it is possible to estimate the ages of the surfaces, but for now the basic result of these age measurements is that the surfaces of the small objects, Mars and the Moon, are much older than those of the large objects, Venus and the Earth, and the geological features on the former were indeed formed billions of years in the past. It is possible to explain this difference using the same kind of dimensional argument that explains why humans can't fly and why elephants have such thick legs.

According to the standard model for the formation of the planets (Chapter 8), the inner planets, when they were first formed, were extremely hot, because of the heat released by the collisions of the smaller objects from which they were assembled. If we assume that their temperatures then were all very similar, the total heat energy within each planet was simply proportional to its volume and hence proportional to its radius cubed (R^3). The energy radiated by a planet is proportional to its surface area (Equation 1.3) and hence to its radius squared (R^2). The time taken for a planet to lose all its initial energy is therefore proportional to $1/R$. Small objects therefore cool faster than big objects, which explains nicely why the surface of our planet is still very active but the Moon is geologically dead.

1.3 Why are planets round?

In the rest of this chapter, I will turn from trying to explain the differences between the planets to trying to find reasons for some of their similarities. The surfaces of the Earth and the other inner planets, for example, are all remarkably smooth – much smoother than an orange although less smooth than a billiard ball. And the biggest similarity of all, which is so easy to take for granted that it is hard to realize that it is an important fact, is that all the planets are spheres. Surprisingly, we can explain both of these facts using a single piece of physics.

The principle of hydrostatic equilibrium is rather obvious once one thinks about it. A planet is a large object with a strong gravitational field, and unless there is something resisting this gravitational field the planet will collapse under its own weight. Since the planets clearly are not collapsing, the principle of hydrostatic

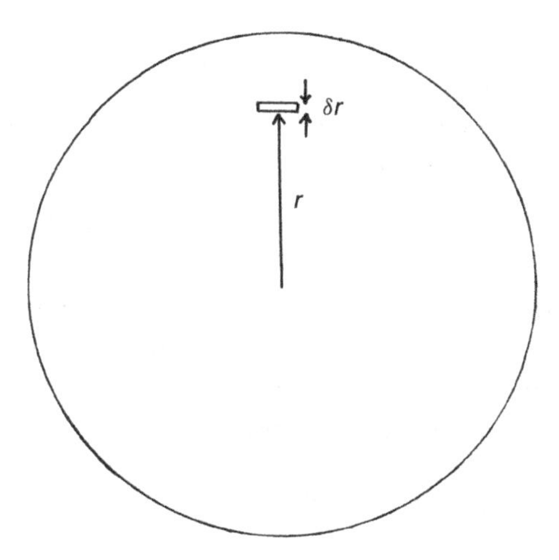

Figure 1.7 A small slab within a planet.

equilibrium states that this inwards gravitational force must be balanced by a pressure gradient within the planet. This is fairly obvious once one considers the forces on a small slab of matter within the planet (Figure 1.7). The material closer to the centre of the planet exerts a gravitational force downwards on the slab (the material further from the centre does not exert a net gravitational force on the slab). The pressure of the material under the slab pushes it upwards, and the pressure of the material above the slab pushes it downwards. If the slab is to stay at rest – to remain in equilibrium – to balance the downwards gravitational force, the pressure below the slab must be slightly higher than the pressure above the slab. The pressure must therefore increase with increasing depth – otherwise the planet would collapse. We can now turn this simple argument into an equation relating the pressure gradient and the density within the planet.

Those without calculus should skip to Equation 1.14, which gives a relationship between pressure, P, and the distance, r, from the centre of the planet, which is derived from the principle of hydrostatic equilibrium. In deriving this relationship, I have had to make one additional assumption: that the density, ρ, does not vary within the planet. Although this assumption is clearly not completely correct – the density of rock at the surface of the Earth ($\approx 3000\,\mathrm{kg\,m^{-3}}$) is lower than the average density of the Earth ($\approx 5000\,\mathrm{kg\,m^{-3}}$) – the equation derived from it does give a fairly accurate picture of how the pressure varies within the Earth. The other terms in the equation are R_p, the radius of the planet, and G, the gravitational constant.

Let us assume that the slab is at a distance r from the centre of a planet and has an area A and thickness δr. We will assume that the planet is a perfect sphere, so all its properties, such as density, ρ, and pressure, P, depend only on r. The volume of the slab is $A\delta r$ and its mass is $\rho A\delta r$. Now let us consider the forces on the slab. The fact that only the material below the slab exerts a net gravitational force on it follows from Newton's law of gravity, although it is not trivial to prove. From Newton's law, the downwards gravitational force on the slab is thus

$$F_{\mathrm{g}} = \frac{GM(<r)M_{\mathrm{slab}}}{r^2} = \frac{GM(<r)A\rho\delta r}{r^2} \tag{1.7}$$

The pressure of the material below the slab will exert an upwards force equal to the pressure times the area of the slab: PA. There will be an additional downwards force from the pressure of the material above the slab: $P(r + \delta r)A$. As the slab is in equilibrium, the sum of the forces must be zero:

$$P(r)A - P(r + \delta r)A - \frac{GM(<r)A\rho\delta r}{r^2} = 0 \tag{1.8}$$

After some rearranging, this equation becomes

$$\delta P = -\frac{GM(<r)\rho\delta r}{r^2}$$

which with a little bit of further arranging becomes

$$\frac{\delta P}{\delta r} = -\frac{GM(<r)\rho}{r^2} \tag{1.9}$$

We can now take the fundamental step of calculus (also incidentally invented by Newton) and allow the thickness of the slab to tend to zero, which yields the basic equation of hydrostatic equilibrium for a sphere:

$$\frac{\mathrm{d}P}{\mathrm{d}r} = -\frac{GM(<r)\rho}{r^2} \tag{1.10}$$

It would be nice now to solve this differential equation to see how the pressure changes with radius within a planet. We can do this, but not without making a simplifying assumption.

Let us think of the planet as being composed of a large number of spherical shells, each of thickness δr. $M(<r)$ is the sum of the masses of the shells interior to r. The mass of a single shell is approximately $4\pi r^2\rho\delta r$. When the thickness of the shells is allowed to tend to zero, the sum of the masses of the shells becomes the integral

$$M(<r) = \int 4\pi r^2 \rho \mathrm{d}r \tag{1.11}$$

The only way to proceed further – without any additional information about the interior of the planet – is to make some assumption about how density depends on radius. One obvious one to try is to assume the density is independent of radius. This is probably not too bad an assumption for solid objects like the Earth. The density of rock at the surface of the Earth ($\approx$3000 kg m^{-3}) is less than the average density of the Earth ($\approx$5000 kg m^{-3}), so the assumption *is* wrong, but we may hope r is dist. from planet center

that it is not so wrong that any conclusion we draw will be invalidated. With this assumption, $M(<r) = 4/3\pi r^3\rho$, and Equation 1.10 becomes

$$\frac{dP}{dr} = -\frac{G4\pi\rho^2 r}{3} \tag{1.12}$$

We can solve this simple differential equation by separating the variables (for those not familiar with this technique, differentiate the solution (1.14) to check that you recover 1.12):

$$\int_0^P dP = -\int_{R_p}^{r} \frac{G4\pi\rho^2 r dr}{3} \tag{1.13}$$

In this equation, R_p is the radius of the planet and I have assumed the pressure at the surface is zero. The solution to the integral is

$$P = \frac{G2\pi\rho^2}{3}(R_p^2 - r^2) \tag{1.14}$$

This equation shows that the pressure increases rapidly with increasing depth, reaching a maximum at the planet's centre. Detailed modelling of the structure of the Earth (Chapter 4) shows the density increases by a factor of only ≈ 3 from the surface down to the centre, whereas the pressure increases much more rapidly, so our assumption that the density does not change at all is probably not misleading. The pressure at the centre ($r = 0$) of the planet is therefore

$$P_{cen} = \frac{G2\pi\rho^2 R_p^2}{3} \tag{1.15}$$

We often use rock as a metaphor for strength – rock-like, granite-faced – but given enough pressure even a rock will be overwhelmed and the chemical bonds between the molecules that give a rock its rigidity and shape will be broken. This critical pressure is about $10^9\,\mathrm{N\,m^{-2}}$. Equation 1.15 shows that at the centre of the Earth the pressure is $1.7\times10^{11}\,\mathrm{N\,m^{-2}}$ – much greater than this critical pressure. Deep inside the Earth, therefore, the metaphor breaks down, and rock behaves more like a liquid than the rigid substance we are familiar with. The equation implies the inner planets can be divided into two distinct regions. At depths less than a critical depth, on the Earth ≈ 25 km, the pressure is less than the strength of rock, and rock behaves like the rigid stuff we are used to; at greater depths, the rock will gradually flow wherever there is a pressure gradient. The former region, where rock behaves like rock, is called the *lithosphere*, the latter region the *asthenosphere*. Seismic observations and more detailed modelling (Chapter 4) imply that the true thickness of the Earth's lithosphere is about 100 km.

This calculation has some interesting implications. First, it shows why the Earth is round. Because rock is able to flow throughout most of the body of the Earth, our planet's shape should be the one with the lowest possible gravitational potential energy – in the same way that water, whenever it has the chance, runs downhill to a position of lower gravitational energy. It is possible to show this shape is a sphere by a simple thought experiment. Suppose we start with a planet that is a perfect sphere and we dig a small hole and pile the dirt by the side of the hole. The gravitational potential energy of the dirt has increased because it has moved up through the gravitational field of the planet, which means the total gravitational energy of the planet must also have increased. Thus anything we do to change the shape of a sphere will increase its gravitational energy. It is therefore not surprising that planets are round.

Small objects in the solar system, however, are not usually round (Figure 1.5), and Equation 1.15 shows why this is so. The pressure at the centre of an object increases as the square of its radius. If the object is small enough, the pressure at its centre will not be greater than the critical pressure for rock. The equation shows that, for a density of 5000 kg m^{-3}, the average density of the Earth, the threshold radius is 535 km. Objects smaller than this, such as Ida, may have any shape because the pressure is not great enough to break the chemical bonds within the rock. The true threshold radius depends on the density of the object and also on its internal structure and composition, which means that in practice the boundary between round and non-round objects is rather fuzzy. This is shown by the case of the largest asteroid, Ceres. Observations with the Hubble Space Telescope show that Ceres is spherical (Figure 1.8), although its radius (475 km) is slightly below the threshold I have calculated.

The last of the common planetary properties we will consider is why planets are rougher than snooker balls but smoother than oranges. The largest mountain

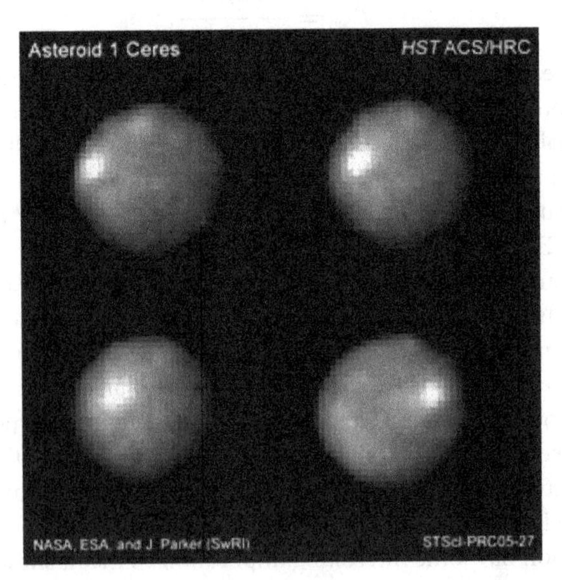

Figure 1.8 Four images of the asteroid Ceres taken by the Hubble Space Telescope over a period of a few hours. The movement of the bright spot is caused by the rotation of the asteroid, which takes about 9 hours (courtesy: J. Parker *et al.* and NASA).

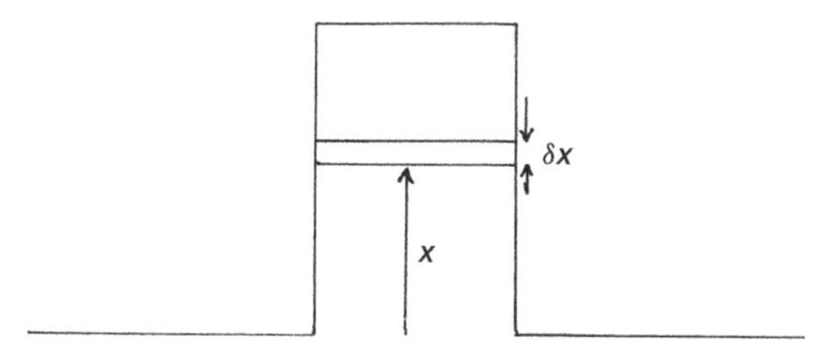

Figure 1.9 An idealized mountain. The two horizontal lines show a slab of thickness δx and height x.

on the Earth is Mauna Loa, which measured from its base, buried deep under the Pacific, has a height of ≈9 km. This is only about 0.15 % of the radius of the Earth, so although mountains look large to us, the Earth is actually remarkably smooth. What determines the size of the wrinkles on the surfaces of the planets?

Figure 1.9 shows a rather unrealistic mountain, which I have represented as a rectangular block (another interesting thing we could consider, but which I do not have space for here, is why mountains have the shapes that they do). As in the case of a whole planet, we can show there must be a pressure gradient within the mountain by considering the forces on a slab of material within it. There is a downwards gravitational force on the slab, and to balance this force the pressure of the material below the slab, which is pushing up on it, must be slightly greater than the pressure of the material above the slab pushing down. We can again use calculus to turn the principle of hydrostatic equilibrium into a relationship between pressure and depth

Those without calculus should skip to Equation 1.20, which gives the relationship between the pressure, P, at the base of the mountain and the height of the mountain, h. The other terms in the equation are M_P and R_P, the mass and radius of the planet, ρ, the density of the planet, and G, the gravitational constant.

The thickness of the slab is δx and the area of the slab is A. The mass of the slab is thus $\rho A \delta x$, in which ρ is the density of the rock. From Newton's law, the gravitational force acting downwards on the slab is

$$F_g = \frac{GM_P \rho A \delta x}{(R_P + x)^2} \tag{1.16}$$

In this equation, M_P and R_P are the mass and the radius of the planet, and I have assumed that the mountain itself does not exert a significant gravitational force on the slab. The upwards force from the pressure of the material below the slab is $P(x)A$; the downwards force from the pressure of the material above is $P(x + \delta x)A$. As the slab is in equilibrium, the sum of the forces must be zero, which gives

$$\frac{GM_P \rho A \delta x}{(R_P + x)^2} + P(x + \delta x)A - P(x)A = 0 \tag{1.17}$$

On the Earth at least, the heights of the mountains are much less than the radius of the planet, so I will make the additional assumption that the x in the denominator of the left-hand term is negligible. After some rearranging and the standard calculus trick of allowing δx to tend to zero, the equation becomes

$$\frac{dP}{dx} = -\frac{GM_P\rho}{R_P^2} \tag{1.18}$$

We can solve this equation by again separating the variables:

$$\int_0^P dP = -\frac{GM_P\rho}{R_P^2}\int_h^x dx \tag{1.19}$$

in which I have assumed that the pressure is zero at the top of the mountain. The pressure at the bottom of the mountain ($x = 0$) is then

$$P = \frac{GM_P\rho h}{R_P^2} \tag{1.20}$$

The mountain will only stand up if this pressure is less than the critical pressure of rock – anything higher and the rock will no longer be rigid and the mountain will be resting on soggy foundations; the rock at the bottom will be squeezed like toothpaste by the weight of the rock above. We can estimate the maximum height of the mountains on a planet by rearranging the equation and replacing the pressure by the critical pressure, P_C:

$$h_{max} = \frac{R_P^2 P_C}{GM_P\rho} \tag{1.21}$$

The equation gives a maximum height for the Earth's mountains of $\approx$30 km, a few times higher than the observed value but not too bad agreement given the simplicity of the calculation.

This equation also allows us to estimate the typical size of the mountains on other planets. The mass of a planet is

$$M_P = \frac{4\pi\rho R_P^3}{3} \tag{1.22}$$

In this equation, the density, ρ, is the average density of the planet, whereas the density in Equation 1.21 is the density of rock in the mountain. For simplicity we will assume these are the same, which makes it possible to combine the two equations:

$$h_{max} = \frac{3P_C}{4\pi G\rho^2 R_P} \tag{1.23}$$

This equation shows that the maximum height of a planet's mountains depends inversely on the radius of the planet. The radius of Mars is a factor of $\approx$2 less than the Earth, which means the maximum height of its mountains should be $\approx$2 times larger. The ratio of the heights of Olympus Mons and Mauna Loa is 2.8, which, given the simplicity of the model, is in reasonable agreement with the prediction. The biggest mountain in the solar system is on Mars because it is a small planet.

1.4 When is a planet not a planet?

I will finish this chapter with a story that includes astronomy, human nature, politics and also the principle of hydrostatic equilibrium. Until recently the term 'planet', like the geographical term 'continent', did not have a precise scientific definition, but the discovery in 2005 of an object in the EK belt that is bigger than Pluto forced the International Astronomical Union (IAU) to invent one. Should this new object, Eris, be considered a planet – in which case the solar system would have 10 planets – or should it not be considered a planet – and if so, Pluto should also clearly not be considered a planet, and the solar system would have only eight planets.

Since Tombaugh discovered it in the 1930s, Pluto has always been the planetary misfit. It is on the outskirts of the solar system, yet is a tiny solid object rather than a gas giant. It also has a very eccentric orbit compared with the other planets (it is sometimes closer to the Sun than Neptune) and a very low mass, much less than the masses of the other planets and only one sixth the mass of the Moon.

Astronomers started to become suspicious that Pluto was not really a planet in 1992 when the EK belt was discovered. Some of the objects in the EK belt have very similar orbits to Pluto. Pluto is in a 3-to-2 orbital resonance with Neptune (Chapter 6), which means it orbits the Sun twice in the time it takes Neptune to orbit the Sun three times, keeping it safe from the gravitational effect of the larger planet. Astronomers soon discovered that some of the objects in the EK belt are in this same orbital resonance – objects for which the term 'plutinos' was quickly coined. The discovery of other small objects beyond the orbit of Neptune, many of which have the same kind of orbit as Pluto, gave rise to the uncomfortable suspicion that Pluto was not really a planet. For several years this remained just a suspicion for two reasons: Pluto was much bigger than the other trans-Neptunian objects and, uniquely, it had a moon, Charon.

But the suspicion began to harden into something more definite in 2004 with the discovery of an object in the EK belt, Sedna, with a diameter of about 1000 km – almost half that of Pluto. Moreover, by now astronomers knew that Pluto was not unique in having a moon. More than 10 objects in the EK belt are now known to have tiny moons. Finally, in July 2005, astronomers at the California Institute of Technology announced they had discovered an object even bigger than Pluto–Eris.

The IAU is the international organization of professional astronomers and, in response to the discovery of Eris, it set up a committee to frame a definition of what we mean by a planet. After careful consideration, the Planet Definition Committee proposed that an object should be considered a planet if it orbits the Sun and is large enough that, as we discussed above, its weight shapes it into a sphere. By this definition, the solar system has 12 planets: the traditional eight, Pluto, Eris, Ceres, and, for a rather technical reason, Pluto's moon, Charon. The IAU executive put forward the committee's proposal as a resolution before the IAU general assembly, which met in Prague in August 2006.

The proposal raised a storm of controversy throughout the astronomy world and, embarrassingly for the committee, the general assembly voted down the resolution by an overwhelming majority. It then approved the alternative resolution that an object is a planet if it satisfies three conditions: (i) it must orbit the Sun; (ii) it must be large enough that its weight shapes it into a sphere; (iii) it must be much larger than any object in its orbital neighbourhood. Pluto, Eris, Ceres and Charon satisfy the first two conditions but not the third, and thus the solar system now has eight planets. As a sop to the defenders of Pluto, the IAU invented a new class – dwarf planet – for objects that satisfy the first two conditions but not the third, but this does not change the basic conclusion: the solar system now has eight planets.

Why did this rebellion occur? I think it was partly because the committee worried too much that people would be upset that an object that had been a planet for 70 years was suddenly not a planet and partly because they were not experts in classification systems. As physical scientists, used to looking for simplicity in nature, the committee members were attracted to the elegant idea of using the principle of hydrostatic equilibrium to define what we mean by a planet. Biologists are much more experienced in classifying things because the relationships between different species is such an important part of biology. A classification system that put cows and horned toads in the same class merely because they both have horns would not be very useful, because everything we know about them – their structures, metabolisms and positions in the evolutionary tree – implies they are very different beasts. In the solar system, the underlying reality is there are eight large objects and two belts of smaller objects. This configuration must have arisen when the solar system was formed, and thus Pluto is more likely to be similar, in early history and composition, to the other objects in the EK belt than to the large objects in the solar system. The classification system proposed by the committee did not reflect this reality, lumping several objects in the belts in with the large objects. Most astronomers instinctively realized this, which is why they voted down the proposal.

The committee also failed to take an opportunity to show the public the true meaning of science. The biggest misconception of science is that it is just a collection of facts. The truth is that science is a powerful method of finding out about the world that is easy to understand and available to anyone. Lists of facts are simply provisional conclusions about the world, which may turn out to be wrong. The fundamental law for scientists – often hard to live up to in practice – is always to be prepared to admit mistakes. The committee missed an opportunity to demonstrate this on the largest possible public stage. Tombaugh's discovery of the ninth planet was a provisional conclusion which, 76 years later, turned out to be wrong.

Exercises

1 The Oort cloud is a cloud of 'dirty icebergs' surrounding the solar system which is believed to be the source of the long-period comets (Chapter 7). The radius of the Oort cloud is about 50 000 AU. Estimate the temperature of one of the objects in the Oort cloud.

2 Use the principle of hydrostatic equilibrium to calculate the thickness of Mars' crust. You should make the assumption that the density of Mars is independent of depth. Use your answer to suggest a possible explanation of why plate tectonics occurs on the Earth but not on Mars.
(Radius of Mars: 3397 km; mean density: 3393 kg $m^{-3)}$

3 (calculus required) Sometime in the far future a strange object enters the solar system. The object is perfectly spherical, completely smooth, has a radius of 500 km and a mass of 2.04×10^{21} kg. Astronauts land on the object and find that the surface is made of iron. Scientists speculate that the object may be a giant spaceship and be hollow inside. Using a value for the density of iron of 8000 kg m^{-3}, calculate the radius of the cavity. Check this hypothesis by using the principle of hydrostatic equilibrium to determine whether the pressure in the iron shell exceeds at any point the tensile strength of iron ($\approx 10^{10}$ N m^{-2}).

2

Other planetary systems

The Galactic Empire was falling. It was a colossal Empire, stretching across millions of worlds from arm-end to arm-end of the mighty double-spiral that was the Milky Way. Its fall was colossal, too – and a long one, for it had a long way to go.

Isaac Asimov (Foundation and Empire)

2.1 The discovery of exoplanets

It now seems that the Galaxy is full of planetary systems, but it is not surprising that it took so long to detect the second planetary system (our own being the first, of course). Planets around other stars are very faint because they are so far away, and they are also very close to a bright object, which makes looking for them rather like looking for a grain of sand in the direction of a bright car headlight.

As an example of how difficult it is to detect planets, let us suppose there is a planetary system exactly like our own but at a distance of 10 parsec. The angle on the sky between the star and the equivalent of Jupiter in this planetary system would be about 0.5 arcsec, which does not sound too bad because the angular resolution of the Hubble Space Telescope (HST) is $\approx$0.1 arcsec, suggesting it should be quite easy to distinguish the planet from the star. The problem arises from the huge difference in brightness between the planet and the star. The luminosity of a sphere, if it radiates like a blackbody, is equal to its surface area multiplied by the power emitted per square metre of the surface, which from Stefan's law is σT^4. The luminosity of both a planet and a star is therefore proportional to R^2T^4. Planets are therefore much fainter than stars for two reasons: they are smaller and cooler. At optical wavelengths, the Jupiter in this system would be about 1 billion times fainter than the star. The angular resolution of the HST would help here

Planets and Planetary Systems Stephen Eales

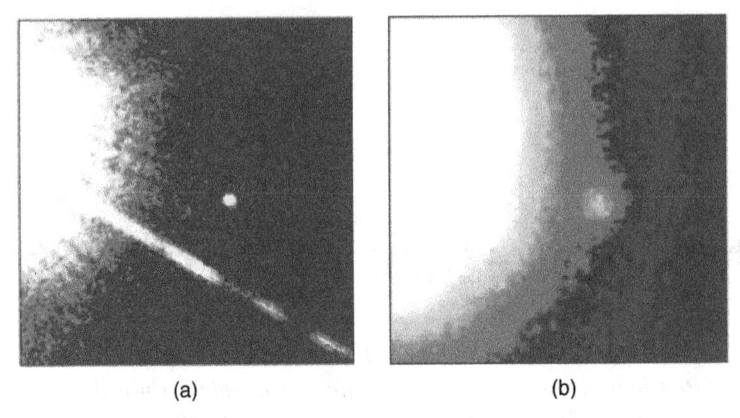

Figure 2.1 Two images of a binary star system. The two stars are about 40 AU apart, the brighter one being about 60 times less luminous than the Sun and the fainter one a brown dwarf with a mass only 30 times that of Jupiter. The brown dwarf is hardly visible in the image in (b), which was taken with a telescope on the ground, but is clearly visible in the Hubble Space Telescope image in (a). In a planetary system like the one described in the text the planet would be 2000 times fainter than the brown dwarf and eight times closer to the star, illustrating the difficulty of seeing a planet against the glare of a star. (courtesy: T. Nakajima, S. Durrance, S. Kulkarni, D. Golimowski and NASA).

but not very much. The angular resolution of a telescope may be measured from a point source, such as a star, and the value usually quoted is the angular distance between the two points on either side of the peak at which the brightness has fallen by a factor of two from its central value, which for the HST is 0.1 arcsec. If the planet is 0.5 arcsec from the star and if it were being observed by the HST, the brightness of the star would therefore have fallen by much more than a factor of two from its central value, but it would still swamp the planet's light. Figure 2.1 shows how hard it is even to detect the fainter stars in binary systems. Thus the obvious method of looking for planets around another star – simply taking a picture – was not the one used to detect the first exoplanet. Before we turn to the method that was successful, note that we would have more chance of taking a successful picture if we changed wavebands. Wien's law states that λ_{max}, the wavelength at which the emission from a black body peaks, is proportional to T^{-1}, and so whereas the emission from a star peaks in the optical waveband, the emission from a planet is at a maximum somewhere in the infrared. At a wavelength of 30 μm, for the planetary system in this example, Jupiter would only be 10 000 times fainter than the star. For this reason, as I will describe later in this chapter, future telescopes whose main goal is to observe exoplanets will work in the infrared.

The method that led to the discovery of the first exoplanet is only possible because something we tend to take for granted is actually not quite true. Planets do not orbit

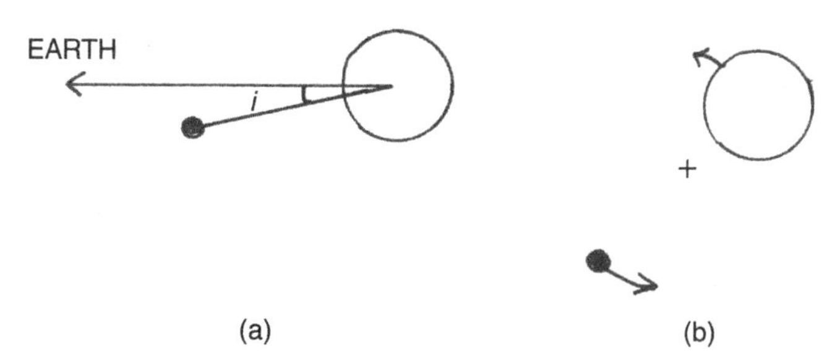

Figure 2.2 Two views of a planetary system. (b) shows the planetary system viewed from above. Both the star and the planet orbit the centre-of-mass, which is marked by the cross. (a) shows the planetary system viewed from the side. If the plane of the planetary system is perpendicular to the line between the star and the Earth ($i = 90^\circ$), the Doppler method will not work.

around their stars, but instead both the planets and the star orbit a point called the *centre-of-mass*. The centre-of-mass of the solar system is just above the surface of the Sun, so the Sun is actually moving in a tiny orbit of its own, caused by the combined gravitational force of all the planets. Any star that is surrounded by planets must be moving in a small orbit around the centre-of-mass of the planetary system, and so one way to look for planets around a star would be to monitor closely the star's position on the sky, which must change as the star orbits the centre-of-mass. Unfortunately, this method doesn't work either, because this change is too small to detect with current instruments. The successful method was not to look for changes in the star's position, but to look for changes in its velocity.

Figure 2.2 shows a very simple planetary system containing only a single planet. Both the planet and the star orbit around the centre-of-mass, and as the star moves, its velocity relative to the Earth changes as it first moves towards the Earth and then away from it. It is possible to measure the velocity of a star from the Doppler shift of its spectral lines: the lines will be shifted to shorter wavelengths (a blue shift) when the star is moving towards the Earth and to longer wavelengths (a red shift) when it is moving away from the Earth. The basis of the method is to look for an oscillation in the wavelengths of a star's spectral lines caused by its orbit around the centre-of-mass. The figure shows that the method will not work in every situation. If the plane of the planetary system is perpendicular to the line between the Earth and the star, the star's velocity relative to the Earth will not change and it will be impossible to tell it has a planet.

Let us now estimate the size of this effect and again see whether it would be possible to use this method to detect a planetary system like our own (you may guess the answer, but my guess is that your guess will be wrong). The centre-of-mass is defined as the average of the mass-weighted positions of the objects in the system. We will assume there is only a single planet in the system, which is a reasonable assumption because Jupiter is so much larger than the other planets. The relationship between the star's distance from the centre-of-mass, x, and the planet's distance, r, is therefore:

$$M_p r = M_s x \tag{2.1}$$

As the planet moves around the centre-of-mass, the centripetal force on it is produced by the star's gravitational force:

$$\omega^2 r M_p = \frac{G M_p M_s}{(r + x)^2} \tag{2.2}$$

The angular velocity, ω, is the same for both the planet and the star – they are always on opposite sides of the centre-of-mass – and is related to the orbital period, T, by

$$\omega = \frac{2\pi}{T} \tag{2.3}$$

The star is much closer to the centre-of-mass than the planet, and so x in Equation 2.2 can be neglected and the equation rearranged:

$$\omega = \left(\frac{G M_s}{r^3}\right)^{\frac{1}{2}} \tag{2.4}$$

The star's velocity, v, is equal to its angular velocity multiplied by its distance from the centre-of-mass. From Equations 2.1 and 2.4, with some rearranging, this is given by

$$v = \omega x = \left(\frac{G}{M_s r}\right)^{\frac{1}{2}} M_p \tag{2.5}$$

As the star moves around the centre-of-mass, its spectral lines will move backwards and forwards in wavelength because of the Doppler shift. The maximum change in wavelength, $\Delta\lambda$, is related to the star's velocity by

$$\frac{\Delta\lambda}{\lambda} = \frac{v \cos i}{c} \tag{2.6}$$

in which i is the angle between the orbital plane of the planetary system and the line joining it to the Earth (Figure 2.2). For most planetary systems, although not for all (see below), we do not know the inclination of the orbital plane to the line-of-sight, which means there is some inevitable uncertainty in the mass estimates of planets made using this method. However, for the moment, let us forget about this complication and assume we can measure the star's velocity precisely from the change in wavelength of its spectral lines. On the right-hand side of Equation 2.5 there are three variables: the planet's mass, the mass of the star and the planet's distance from the centre-of-mass. It is usually possible to estimate the star's mass from its spectral type, which however still leaves two unknowns in the equation. However, we can also measure the time taken for the spectral lines to oscillate backwards and forwards in wavelength, which tells us the orbital period of both the star and the planet and also, through Equation 2.3, the angular velocity of the two objects. The planet's distance from the centre-of-mass can be written in terms of the orbital period using Equations 2.4 and 2.3.

$$r = \left(\frac{GM_s}{\omega^2}\right)^{\frac{1}{3}} = \left(\frac{T^2 GM_s}{4\pi^2}\right)^{\frac{1}{3}} \tag{2.7}$$

Equations 2.5 and 2.7 can now be combined to produce an equation linking the thing we would like to estimate – the mass of the planet – with the two things which can be measured from the spectral lines: the orbital period and the velocity of the star:

$$M_p = v\left(\frac{M_s^4 T^2}{G^2 4\pi^2}\right)^{\frac{1}{6}} \tag{2.8}$$

One nice thing about this equation is that it is independent of the distance to the star – the method should be just as good at finding planetary systems on the other side of the Galaxy as it is at finding ones close to the Sun.

Using this equation, I calculate that the amplitude of the velocity oscillation produced in the star by the 'Jupiter' in our hypothetical planetary system would be approximately 12 m s^{-1}. This is barely detectable with current technology and was certainly not detectable in the 1990s when the first exoplanets were discovered. This is why I suspected your guess would be wrong. But if planetary systems like our own are not that easy to detect using this method, why have so many other planetary systems been discovered?

The first planet discovered outside the solar system orbits the star 51 Pegasi. This star lies at a distance of about 16 parsec and is fairly similar in luminosity and colour to the Sun. In 1995, two Swiss astronomers, Michel Mayor and Didier Queloz, started monitoring the wavelengths of the star's spectral lines and discovered that the lines are oscillating backwards and forwards in wavelength every four days. The amplitude of the velocity oscillation is about 50 m s^{-1} (Figure 2.3). From the period and amplitude of the oscillation, Mayor and Queloz were able to use Equations 2.7 and 2.8 to estimate the planet's mass and its distance from the star. As I said above, there is some uncertainty in these values because we do not know the orientation of this planetary system, but with this caveat the planet's mass is about 0.5 times the mass of Jupiter and its distance from 51 Pegasi is about 0.05 AU.

As I write (September 2008), there are 228 exoplanets known and the number is increasing all the time. Figure 2.4 shows the masses of these planets plotted against the distances from their stars. The figure shows that these planetary systems are

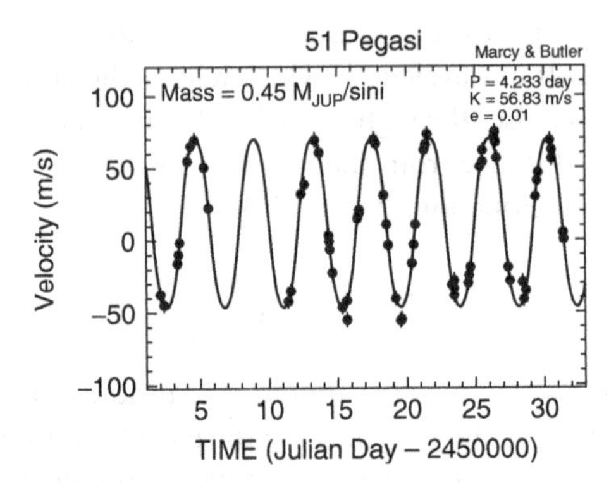

Figure 2.3 The variation in the velocity of the star 51 Pegasi caused by an unseen planet (courtesy: Geoff Marcy).

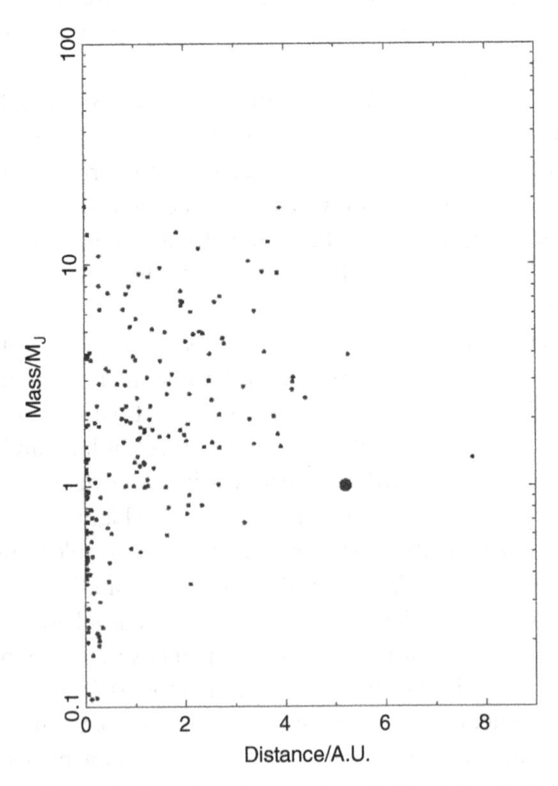

Figure 2.4 Plot of the masses of known exoplanets (in units of the mass of Jupiter) against their distances from their stars. The big circle shows Jupiter itself.

very different from the solar system because the giant planets in these systems are generally very close to their stars; in about 40 % of cases the planet is closer to the star than Mercury is to the Sun. Astronomers have been quite lucky. Equation 2.5 shows that the size of the Doppler effect increases as the distance between the planet and the star decreases. If these 'hot Jupiters' did not exist and if other planetary systems were like our own, we would probably still be waiting for the discovery of the first exoplanet.

2.2 The implications of the existence of other planetary systems

There are several conclusions one can draw from this discovery of a large number of planetary systems. The first is so obvious that it may seem of little value, but it is actually very important: making planets is easy. Consider the situation before 1995. We then knew of only one planetary system, and so it was impossible to know whether planet formation was something that always happened when a star formed or whether it was something that only occurred extremely rarely. The fact that planetary systems are quite common shows that any theory that suggests the opposite, such as the old theory that the solar system was formed from a filament of gas pulled out from the Sun by the gravitational field of a passing star, can be immediately ruled out.

It also has implications for one of the biggest questions of all: whether there is life elsewhere in the universe. In the 1960s the astronomer Frank Drake made a first attempt to estimate the number of extraterrestrial civilizations in the Galaxy (Chapter 9). As you might expect, Drake's estimate depended on a large number of unknown factors, and one of these was the probability of a star having planets. This probability was very uncertain because we then knew of only one planetary system. It was not possible to use the existence of the solar system to infer that planetary systems must be quite common, because suppose that the formation of planetary systems is so rare that it has happened only once in the history of the universe – we would inevitably find ourselves the winners of the cosmic lottery, the lucky inhabitants of that solitary planetary system. Although most of the factors in Drake's equation are still extremely uncertain (if not completely unknown), at least we now know the probability of a star having planets is actually quite high.

The third conclusion one can draw, although this is less obvious, is that our knowledge of how planets form is incomplete. The standard theory, first suggested by the Marquis de Laplace two centuries ago, is that planets are formed out of rotating discs of gas and dust around a newly formed star (Chapter 8). This theory can plausibly account for why the giant planets in the solar system are a long way from the Sun, because the heat from the newly formed Sun stopped compounds with a low melting point solidifying in the inner part of the disc, which meant there was less material there to act as the nucleus of a giant planet. The discovery of hot Jupiters was therefore a surprise. As I describe in Chapter 8, one possible explanation is 'planetary migration' – the idea that a hot Jupiter may have formed further out in a protoplanetary disc and then gradually moved inwards as the result of its gravitational interactions with the small objects left over in the disc after the formation of the planets.

There is another obvious conclusion one could draw, but this one is actually incorrect. It is tempting to conclude, because most of the planetary systems so far discovered contain giant planets close to the star, that the solar system must be a rather unusual planetary system. However, with current technology, small planets like the Earth would be impossible to detect, and even the giant planets in planetary systems like our own would be very difficult to detect. Most of the stars that have been studied using this method do not appear to have planets, and although it is possible that these stars genuinely do not have planetary systems, it seems more likely that many of them do have planets but, as in our own planetary system, the giant planets are further away from the star and so produce a smaller Doppler signal.

The Doppler method of finding planets is an indirect method, and for several years after the discovery of the first exoplanet there remained the possibility there was some other explanation of the oscillating spectral lines. Perhaps the surface of the star was moving up and down – beating like a heart – which would also produce oscillating spectral lines. Five years later, a second method showed conclusively that the oscillating spectral lines *are* produced by planets and not by some strange stellar phenomenon.

This method is also an indirect method and again relies on the effect of the planet on the star, in this case its effect on the star's brightness. Suppose the orientation of a planetary system is such that the planet moves between the star and the Earth (Figure 2.5). As the planet moves between the star and the Earth, it will obscure part of the star's disc, and the star's brightness will fall slightly, returning to normal when the planet is no longer between the Earth and the star. It is easy to calculate the size of this effect. The fraction of the star's light obscured by the planet is the ratio of the areas of their discs, which is just the square of the ratio of their diameters, $(\mathrm{d_p/d_s})^2$. The ratio of the brightness of the star when the planet is in the way to its unobscured brightness is thus $1 - (\mathrm{d_p/d_s})^2$ and the change in its apparent magnitude is

$$\Delta m = -2.5 \log_{10}\left(1 - \frac{d_{\mathrm{p}}^2}{d_{\mathrm{s}}^2}\right) \tag{2.9}$$

For our hypothetical planetary system – one like our own but at a distance of 10 parsec – the change in magnitude would be about 0.01 magnitudes, which is a small effect but one that is possible to detect with careful observations. The difficulty in using this method is that only about 1 % of planetary systems will be at the right orientation such that the planet passes directly between the Earth and the star.

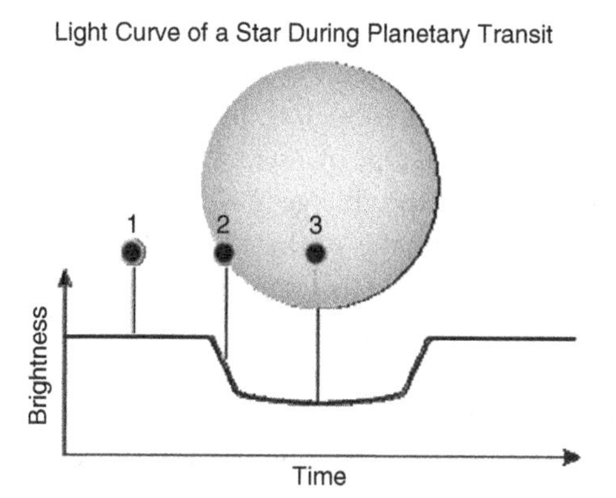

Figure 2.5 Illustration of how a star's brightness changes when a planet passes between the star and the Earth.

In 2000, after monitoring a large number of stars shown by the Doppler method to have planets, a team led by the American astronomer David Charbonneau discovered one whose brightness dips in exactly this way (Figure 2.6). This discovery showed conclusively that the star, HD 209458, really has got a planet. It also allowed the team to find out more about the planet than was possible with the Doppler results alone. For a start, it removed the uncertainty about the inclination of the planetary system, and so the team was able to make more accurate estimates of the planet's mass and its distance from the star. A convenient mass unit in exoplanet research is the mass of Jupiter – M_J. According to the team's calculations, the planet

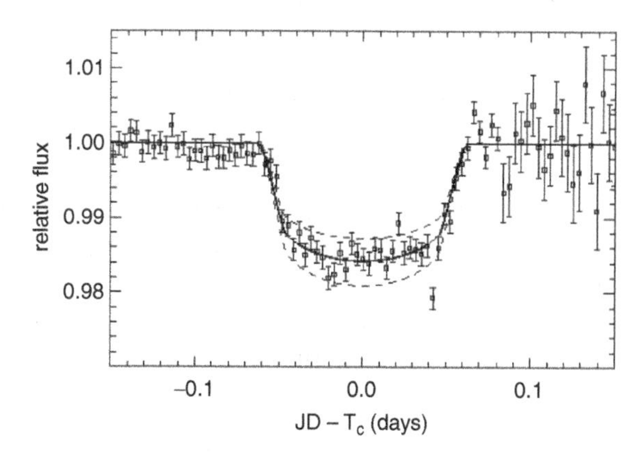

Figure 2.6 Variation in the brightness of the star HD 209458. The brightness drops in exactly the way expected if a planet has moved between us and the star (compare this figure with Figure 2.5), returning to its usual value when the planet is no longer obscuring the star's disc (courtesy: David Charbonneau).

orbiting HD 209458 has a mass of 0.69 M_J and is 0.045 AU from the star, which is only 11 % of the distance between Mercury and the Sun. The predicted temperature of the planet is about 1500 K, hot enough to melt iron, so this planet really is a hot Jupiter. If one knows the spectral type of a star, it is possible to estimate its diameter, and so the team was able to use the change in magnitude when the planet obscured the star to also estimate the diameter of the planet. From the planet's mass and its diameter, they were able to estimate the planet's density. This is 380 kg m^{-3}, which is less than the density of the least dense planet in our own planetary system, Saturn, and is much less than the density of water – so the planet would float if we could only find an ocean large enough.

Since this discovery, astronomers have been very creative in thinking of new ways of investigating this planet and the few other transiting planets that are now known. Several groups have monitored the spectrum of HD 209458 as the planet passes between the star and the Earth. By looking for spectral absorption features that are only present when the planet is between us and the star, they have been able to investigate the planet's atmosphere. They have not got much further than detecting a few of the standard elements – hydrogen, oxygen, carbon and sodium – but they have discovered that the spectral lines often have large Doppler shifts, showing that gas is flowing away from the planet. This outflow of gas is probably explained by the high temperature, which increases the chance of an atom or molecule having enough energy to escape the planet's gravitational field. The best estimate of the outflow rate implies that the planet will eventually lose all of its atmosphere, which has led to a suggestion that it should be called Osiris, after the Egyptian god who lost part of his body to his evil brother Set. Another team has used the Spitzer Space Telescope to measure the combined infrared emission from the planet and star when they are not obscuring each other. By subtracting from this the emission from the star when it is in front of the planet (and so the emission is then only from the star), they claim to have made the first detection of *emission* from an exoplanet; the spectral lines they detect from the planet suggest there may be clouds of silicon in the planet's atmosphere.

Despite these clever ideas, it seems unlikely that we will learn much more with current instruments and telescopes. At the moment, we cannot even detect the giant planets in planetary systems like our own, let alone small rocky planets, which are of great interest partly because we live on one and partly because a small rocky planet is the only one on which we are sure there is life. We would also like to be able to observe exoplanets directly, rather than by the current indirect methods, because this would allow astronomers to use all of their standard forensic tools: imaging, photometry, spectroscopy and polarimetry.

2.3 The future for exoplanet research

Despite its past success, the Doppler method does not hold much promise for the future. A fundamental limit is that the velocities of the stars themselves do vary slightly even if there is no planet; a stellar surface generally moves up and down with a speed of $\approx 1\ \mathrm{m\,s^{-1}}$ due to oscillations within the star. Given this limit, Equation 2.8 shows that it may be possible to find planets in planetary systems like our own with masses a few times less than Jupiter but not planets with masses similar to the inner planets.

The transit method has more promise. A Jupiter-sized planet moving across the face of the star produces a decrease in the star's brightness of ≈ 0.01 magnitudes (Equation 2.9). From the Earth, it is impossible to detect planets that are much smaller than this because photometric accuracy is limited by the Earth's atmosphere. The solution is to go into space. In December 2006, an international consortium led by France launched Corot, a 30-cm space telescope. Corot will monitor the brightness of over 100 000 stars with sufficient accuracy that it should detect even small planets if they pass in front of the stars. Kepler, a similar US mission, will be launched in spring 2009.

Corot and Kepler will answer questions about the statistics of planetary systems (How many are there? Is the solar system unusual?), but they will not tell us much about the planets themselves. The crucial thing here is to be able to observe the planets directly, and the big problem, as I described at the beginning of this chapter, is that planets are faint objects close to very bright objects.

The two vital things necessary to overcome this problem are enough basic sensitivity to detect the planet and enough angular resolution to separate the light of the planet from the light of the star. Astronomers are planning several new telescopes, on the ground and in space, which should (if they work) be able to observe Earth-like planets around other stars.

One solution to the problem is to build bigger telescopes. The European Southern Observatory is planning to build a telescope with a mirror 40 m in diameter, which is about half the length of a football pitch. The sensitivity of a telescope depends on its collecting area, and so the Extremely Large Telescope (ELT) will have much more sensitivity than existing telescopes, easily enough to observe exoplanets. The big uncertainty is whether it will achieve the angular resolution necessary to remove the glare of the star. The best possible angular resolution that can be achieved by a telescope is approximately λ/D radians, in which λ is the wavelength of the radiation being detected and D is the diameter of the telescope. Therefore, in principle, the ELT should have a resolution in the optical waveband of ≈ 0.003 arcsec, approximately 20 times better than the HST. In practice, though, telescopes on the ground are usually limited to a resolution of ≈ 1 arcsec by the effects of atmospheric turbulence. Astronomers have begun to overcome this problem with ground-based telescopes by the technique of *adaptive optics*. This consists of monitoring the atmospheric turbulence by continuous observations of a bright star, and then using this information to correct for the effect of the turbulence by either making small changes to the shape of the telescope's mirror or making changes within the camera optics. This technique has begun to give good results on small telescopes. If it can be successfully used on the ELT, this telescope will have both enough sensitivity and resolution to observe exoplanets.

The alternative is to go into space. Both the European Space Agency (ESA) and NASA have space telescopes on the drawing board whose main goal is to observe small rocky planets around nearby stars. The ESA mission is called Darwin and is tentatively scheduled to be launched in 2015 although, money and politics being what they are, it is quite possible that Darwin and the NASA scheme, the Terrestrial Planet Finder, will eventually be merged into a single mission. The ESA plan is that Darwin will be an interferometer. Interferometers were first invented by radio astronomers to overcome the problem that radio waves have much longer wavelengths than optical light, and so the angular resolution of a single radio telescope is rather poor. The early radio astronomers found that they could achieve much better angular resolution if they combined the signals from several radio dishes. The angular resolution of an interferometer is also given by λ/D, but in this case D is the distance between the dishes rather than the diameter of an individual dish. ESA's current plan is that Darwin will consist of three telescopes, each with a mirror 3 m in diameter, with a separate communications hub to send the signals to the Earth (Figure 2.7).

Darwin will operate at infrared wavelengths because in this waveband the difference in brightness between planets and stars is much less than in the optical waveband. This is the main reason why Darwin will be in space; the Earth's atmosphere is largely opaque to infrared light and the Earth itself is also a strong source of infrared radiation. Astronomers will be able to overcome the problem of

Figure 2.7 An early artist's impression of what Darwin would look like. The current plan is that Darwin will consist of three telescopes and a communications hub (courtesy: ESA).

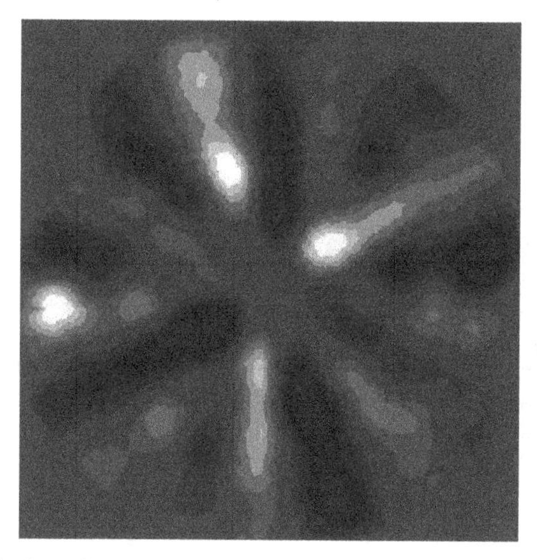

Figure 2.8 Simulation of what Darwin would see if pointed at a planetary system like the solar system at a distance of 10 parsec. The three blobs of light are 'Venus,' the 'Earth' and 'Mars' (Mercury is too faint to see) (courtesy: Bernard Mennesson).

seeing a faint object in the glare of a bright object by playing a clever trick. Darwin will be a *nulling interferometer*, which means that by adding different delays to the signals from the telescopes, astronomers will be able to arrange that the signals from the star interfere destructively, and so the star will no longer be seen. Figure 2.8 shows a simulation of what Darwin would see if pointed at our hypothetical planetary system. The star in the middle has completely vanished; Mercury is a bit too faint to detect; but Venus, the Earth and Mars can all be seen.

The most exciting thing about all these planned telescopes is that they should be able to tell whether there is Earth-like life on the planets they detect. To any extraterrestrials in spacecraft orbiting the Earth at the moment, the biggest clue that there is life on our planet is not some artefact such as the Great Wall of China but something in the atmosphere. The large percentage of oxygen in the atmosphere is a clear sign there is life on Earth, because oxygen is so reactive that if it were not continually replenished by biological activity (photosynthesis in planets), it would rapidly combine with other atmospheric gases and with rocks. Indeed, oxygen is so reactive that the first production of oxygen by the first photosynthetic plants billions of years ago posed a huge problem for life on Earth, and evolution had to generate many new metabolic systems to avoid life being destroyed by the new poisonous gas it was producing.

The extraterrestrials would not actually have to travel to the solar system to discover the existence of life on Earth, because the oxygen in the atmosphere produces spectral absorption lines, and so the extraterrestrials would simply have

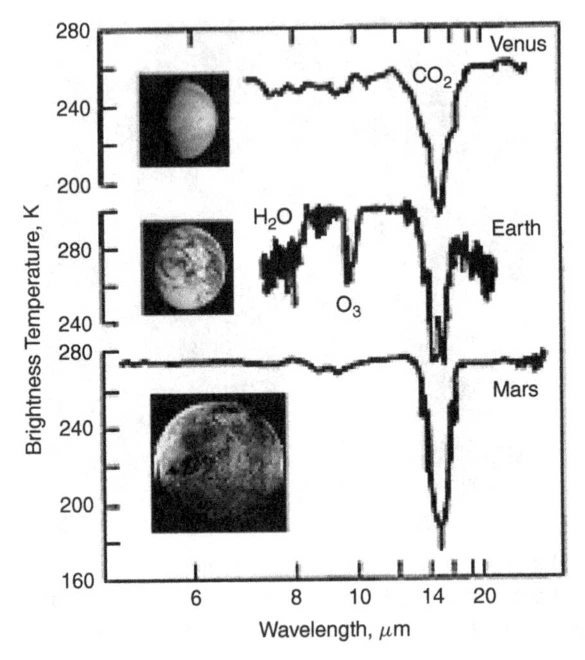

Figure 2.9 Infrared spectra of Venus, the Earth and Mars. Note the absorption feature due to tri-atomic oxygen (ozone) in the Earth's spectrum (courtesy: NASA).

to obtain the Earth's spectrum. Conversely, we will be able to tell that any planet detected by the ELT or Darwin or the Terrestrial Planet Finder contains life without travelling there. There are other spectral 'biomarkers': ozone, the tri-atomic form of oxygen, has a strong spectral line in the infrared waveband (Figure 2.9), and nitrous oxide and methane, which on Earth are also mostly the result of biological activity, also produce spectral lines in the infrared. These last two are not unambiguous biomarkers because these gases can also be produced by other means (methane is produced by volcanoes), but the presence of ozone and oxygen in the atmosphere of an exoplanet would be an unambiguous sign there is life on its surface.

The discovery of other planetary systems has made it seem much more likely that life on Earth is not the result of a huge rollover lottery win and that there is life elsewhere in the universe. With these new telescopes, there is at least a chance that by the end of the next decade we will know for certain whether this is true.

Exercises

1 Suppose that a star has a planet like the Earth at exactly the same distance from the star that the Earth is from the Sun. On the assumption that the star has the same radius as the Sun, calculate the change in the magnitude of the star if the planet passes between the star and the Earth (radius of the Sun: 7×10^8 m; radius of Earth: 6.4×10^6 m).

2 Calculate the variation in the velocity of the star in the first question caused by the gravitational effect of the planet. You should assume that the star has the same mass as the Sun (mass of Earth: 6×10^{24} kg).

3 Suppose that you try to detect this planet by optical imaging. Estimate how much fainter the planet will be than the star. You should make some sensible assumption about the albedo of the planet (radius of Earth: 6.4×10^6 m).

Further Reading and Web Sites

The web sites www.exoplanets.org and www.exoplanet.eu (both accessed 17 September 2008) contain lists of all known exoplanets and are the best places to look for the most recent research.

3

The surfaces of the planets

How clear it is that stones have handled time,
in their fine substance there's the smell of age...
Pablo Neruda

3.1 Rocks

Planetary science is the ultimate interdisciplinary subject. As a physicist, my natural scientific approach is to try to reduce the complexity of the natural world to simple physical laws. As I showed in Chapter 1, this approach can be a very powerful one. It can be used, for example, to show why the planets are round and why the typical size of mountains is different on the different planets. But Chapter 1 also showed this approach has limitations. The solar system is a splendidly diverse place, and there is no way one could start with some simple physical laws and predict the astonishingly different planets that one actually observes. The subject of this chapter is another place where the approach of the physicist breaks down. The surfaces of planets are complicated places, covered in interesting features, continuously changed by processes under the surface, by the atmosphere, by the action of fluids running over the surface and by biological processes – the product of both the overall history of the planet and the individual history of the particular piece of the surface one is looking at. To understand planetary surfaces one needs to follow the geologist's approach of embracing the complexity of the natural world, of investigating the history of individual rocks and of individual geological features, and only occasionally, and very cautiously, drawing general conclusions.

An important tool for anyone trying to understand the history of a lump of rock is a geologist's hammer because the surface of a rock has often been modified by the action of the atmosphere, and it is only by looking inside the rock that one can

Planets and Planetary Systems Stephen Eales

see its true physical structure. Chemical tests are also important for determining the minerals that make up the rock. This kind of rough handling and experimentation in the laboratory has only been possible for rocks on Earth, every mile of which has probably now been walked over by human geologists, and so I will spend the first half of this chapter describing what we know about the surface of the Earth. Only then will I consider the surfaces of the other planets, using as a first guide the knowledge geologists have acquired about the surface of our own. (I will ignore the gas giants, which may well not have surfaces at all.) Although our knowledge of the geology of the other planets is far less than our knowledge of the geology of the Earth, it is worth noting that we do know a little at least about the geology of the Moon because of the 382 kg of rocks brought back by the Apollo astronauts; and if no human geologist has visited Mars, at least there are three robot geologists, complete with tools for probing below the surface of the rocks, currently studying the Martian surface (see below).

There are three main classes of Earth rock. *Igneous* rocks are ones that are formed when magma cools. *Sedimentary* rocks are ones that are formed as the result of the effect of the atmosphere or running water or as the result of biological processes. *Metamorphic* rocks are sedimentary rocks that have been transformed into new types of rock by high temperatures and pressure. We will consider the three types of rock in turn, but first we need to consider what rocks are made of.

Rocks are made of minerals. The particles that are visible when a rock is looked at through a magnifying glass (Figure 3.1) are usually crystals of separate minerals. To complicate things, a mineral is not simply defined by its chemical composition but also by its crystalline structure. The standard example is that of diamond and graphite, which are both composed of carbon but have different structures at the

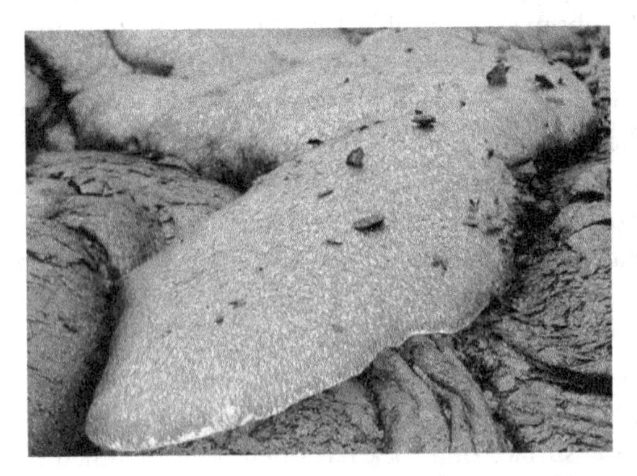

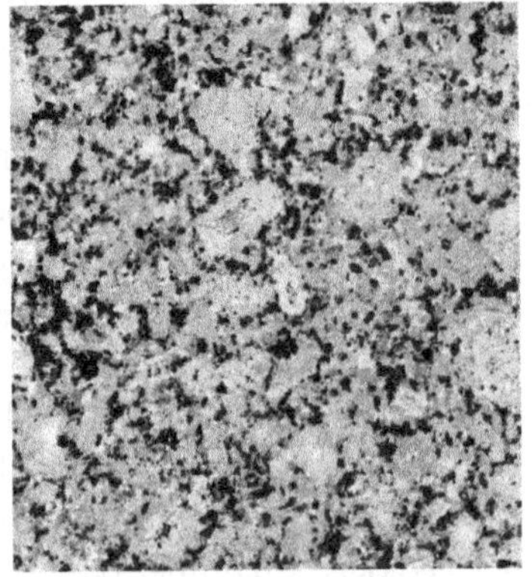

Figure 3.1 Examples of the two main types of igneous rocks. On the left is a basalt rock formed in a lava flow. On the right is common granite. The speckled look of the rock is produced by large crystals of the different minerals making up the granite, which formed as the magma cooled slowly under ground (reproduced courtesy of US Geological Survey).

Table 3.1 Mineral classes.

Class of mineral	Examples
Native elements	Copper, gold
Sulfides and similar compounds	Pyrite – FeS_2
Oxides and hydroxides	Haematite – Fe_2O_3
Halides	Salt – NaCl
Carbonates and similar compounds	Calcite – $CaCO_3$
Sulfates and similar compounds	Barite – $BaSO_4$
Phosphates and similar compounds	Apatite – $Ca_5F(PO_4)_3$
Silicates and similar compounds	Pyroxene – $MgSiO_3$; quartz – SiO_2; olivine – $(Fe,Mg)_2SiO_4$; feldspar – $(K,Na)Si_3O_8$, $CaAl_2Si_2O_8$

molecular level. Table 3.1 lists the main mineral classes. On the Earth and probably on the other inner planets the most important minerals are the silicates, composed of silicon, oxygen and various metals. A particular silicate mineral may have a number of different possible chemical formulae. In the case of olivine, for example, the brackets in the table show that the metal may either be iron or magnesium; feldspars may contain either potassium or sodium. Feldspars and quartz are the lightest of the silicate minerals and are thus expected to rise to the surface in a molten planet – about 60 % of the rock on the Earth's surface is made of feldspars. Olivine and pyroxene are heavier and thus are probably important constituents of the Earth's mantle.

Igneous rocks can be roughly divided into two types based on their mineralogical composition. Granite rocks contain mostly feldspars and quartz, whereas basalts contain mostly heavier silicates such as olivine and pyroxene. There is also generally a difference in where the two types of rock are made. Rocks form from cooling magma, and this can occur either underground, in which case the rocks are called *intrusive or plutonic*, or above ground from a lava flow, in which case the rocks are *extrusive* or *volcanic*. Common granite is a plutonic rock, and the slow cooling underground results in larger mineral crystals (Figure 3.1). Basalts are volcanic rocks and the fast cooling results in much finer grains (Figure 3.1) or even, if the cooling is very fast, in a glassy material. This is not true of every type of granite and basalt. Rhyolite, for example, is a kind of granite that forms out of lava flows on the surface and thus has very small grains. Granites are the most common plutonic rocks found on the Earth's continents. The oceanic crust, on the other hand, is largely made up of basalt. Basalts are undoubtedly common on other objects in the solar system. The analysis of Apollo rock showed that the dark areas on the Moon, the maria (Figure 8.5), are largely made up of basalts. A likely reason why basalts are more prominent on planetary surfaces than granites is the lower viscosity of magma with a basaltic chemical composition, which means a basaltic magma will flow further over a planetary surface than one with a granite composition.

The second type of rock, sedimentary rock, is found only on planets where there is an atmosphere, liquid running over the surface or life. Many sedimentary rocks are produced in a three-stage process, the first stage being the breaking up of igneous rock by mechanical weathering. An example of effective mechanical weathering is when water seeps into a crack in the rock and then freezes; as the water freezes it expands, increasing the size of the crack and gradually breaking up the rock. The product of the weathering depends on the mineral. Quartz, for example, produces sand; feldspars produce clays. The second stage of the process is sorting. A lump of granite may be weathered into a mixture of clay particles and sand grains. Since the sand grains and clay particles have different sizes, the wind and water flowing over the surface will transport them different distances, effectively sorting the clay from the sand. The final stage in the process is when a layer of clay particles or sand grains become compacted and compressed into rock under the weight of the layers above. The sand becomes sandstone. The clay becomes shale.

A particularly important sedimentary rock for life on Earth is limestone. Limestone is produced in two ways, by a chemical process and as the result of biological activity. The chemical process is the *Urey weathering reaction*, in which carbon dioxide dissolved in water reacts with silicates in rocks. One variant of this process is the following two reactions:

$$CaSiO_3 + 2CO_2 + H_2O \rightarrow Ca^{2+} + SiO_2 + 2HCO_3^-$$
$$Ca^{2+} + 2HCO_3^- \rightarrow CaCO_3 + CO_2 + H_2O$$

The calcium carbonate ($CaCO_3$) produced by these reactions is dissolved in the water and is eventually precipitated to form the sedimentary rocks, limestone and chalk. On Earth, however, limestone and chalk are also produced by biological processes. The shells of many tiny marine organisms are made of calcium carbonate. When they die, these creatures sink to the bottom of the oceans, and most limestone is probably the result of the gradual compression of the deposits of these shells.

Limestone and chalk are so important because they are effectively a huge reservoir of carbon dioxide. The amount of carbon in these sedimentary rocks is roughly 20 000 times greater than the amount in all the Earth's deposits of coal and oil. The discrepancy between the amount of carbon dioxide on Venus and the Earth (Chapter 1) is explained by the carbon dioxide locked up in these sedimentary rocks. These rocks are part of a global carbon cycle because when tectonic plates are forced under the Earth (see below) the heat breaks down the calcium carbonate, and carbon dioxide is released back into the atmosphere through volcanic activity.

The final class of rocks are metamorphic rocks. These are sedimentary rocks forced beneath the surface, which are there transformed by the heat and pressure into a different kind of rock. Limestone is transformed into marble, shale into slate.

3.2 Geological structures

Our knowledge of the surfaces of the other planets is still mostly based on images taken by orbiting spacecraft (Chapter 1), and the geologist's understanding of the features on the surface of the Earth has been helpful for planetary scientists when they have tried to interpret these images. If a feature on a Magellan image of Venus, for instance, looks like a well-known geological structure on the Earth, it seems reasonable to assume that the two might have been formed in the same way. This is slightly dangerous, because the conditions on the two planets are very different, so two features that look superficially the same might actually have very different causes, but as our knowledge of the surfaces of Mercury and Venus, in particular, is almost entirely confined to the images taken by Mariner 10, Messenger and Magellan, there is often no alternative.

On the Earth, geological structures can be divided into local ones and structures covering the whole surface. Figure 3.2 shows two examples of a local structure, a fracture in the crust caused by forces within it. On the left is a *reverse fault*, which is caused by forces compressing the crust; the crust fractures and one part of the crust is pushed over the other. On the right is a *normal fault*, which is caused by forces stretching the crust; the crust fractures and parts of it are pushed up and parts pushed down, forming the distinctive *horst* and *graben* feature. A beautiful example of a graben is the Rhine Valley.

The existence of geological structures covering the whole of the Earth's surface was only discovered in the 1950s, although this had been pre-shadowed in the ideas, four decades earlier, of a German meteorologist.

In 1915, in the book *The Origin of Continents and Oceans*, a German meteorologist, Alfred Wegener, proposed that the continents were slowly moving around the Earth. At the time, the idea of *continental drift* must have sounded the kind of crazy idea that professional scientists occasionally receive in letters from members of the public, but it was actually based on some detailed scientific observations. One can explain the

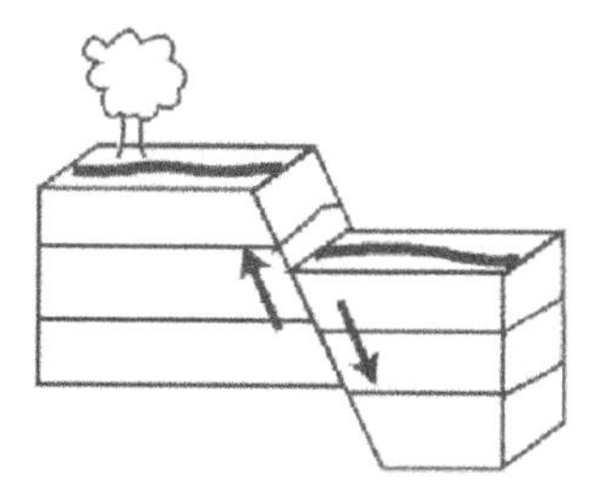

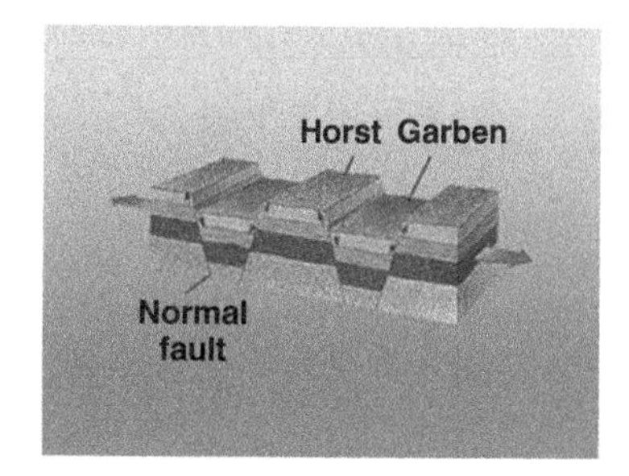

Figure 3.2 Examples of faults, fractures in the Earth's crust. On the left is a reverse fault caused by compressional forces within the crust. On the right is a normal fault caused by forces stretching the crust (reproduced courtesy of US Geological Survey).

differences in the fauna seen on the different continents today – horses in Eurasia, llamas in South America, kangaroos in Australia – by the course that evolution took on the different continents. But Wegener noticed that if one looks in the fossil record the same species is often found in places that are far apart. Fossils of the *mesosaurus*, for example, a small reptile that prowled the Earth 250 million years ago, are found on the west coast of Africa and on the eastern bump of South America that juts into the Atlantic Ocean. In what we can now see was a brilliant insight (people didn't at the time), Wegener also noticed that if one moves the continents around like pieces of a jigsaw puzzle, they actually fit together remarkably well; South America's bump fits neatly into the right-angled corner of Africa's west coast, for example (Figure 3.3). Wegner suggested that the continents had once been part of a single super-continent, which he named Pangaea, which had split apart millions of years ago, with the continents then slowly drifting to their current positions. His idea explained why the *mesosaurus* and other fossils are found in widely separated regions, because if it was true these regions were once connected to each other. Of course, nobody believed Wegener at the time because he could not answer one fundamental question: what process could possibly transport such a heavy thing as a continent?

Alfred Wegener died in 1930 during a scientific expedition across the Greenland ice cap. His idea, although in a rather different form, was resurrected in the 1950s and 1960s as the result of discoveries made by the first intensive surveys of the ocean floor, which until that time was almost as unknown as the far side of the Moon. The first big discovery was of the Mid-Atlantic Ridge, a chain of mountains that rises about 4500 metres above the ocean floor and extends the length of the Atlantic Ocean (Figure 3.4). These mountains are part of a global chain that wends its way

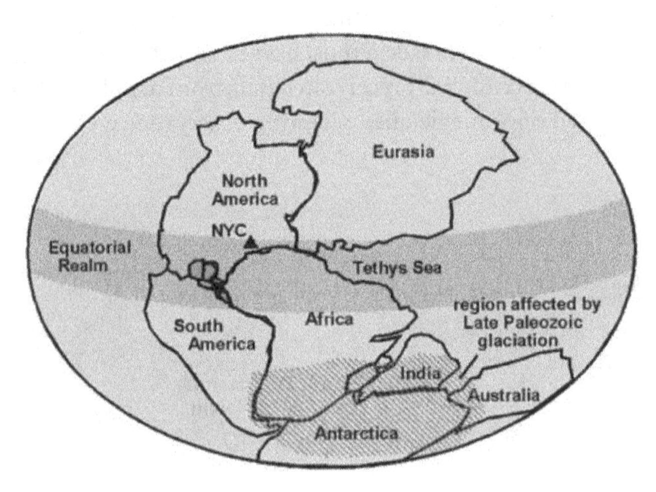

Figure 3.3 The super-continent proposed by Wegener. Geologists now believe that Pangaea broke into separate continents about 180 million years ago (reproduced courtesy of US Geological Survey).

Figure 3.4 A topographic map of the Earth with the oceans emptied of water. Note the ridge of mountains extending along the middle of what was the Atlantic Ocean (reproduced courtesy of University of California Museum of Paleontology).

from ocean to ocean around the globe and is about 50 000 km in length. Viewed from space – and if the oceans were emptied – this chain of mountains would be the most prominent geological structure on Earth.

The second discovery was about the age of the ocean floor. At the time, it was believed that the ocean floor was very old, implying it should be covered by a thick blanket of sedimentary rock, but surveys of the floor of the Atlantic found only a thin layer of this rock. The explanation of why the ocean floor is so young was inspired by the third discovery.

The Earth's magnetic field is caused by the motion of the iron in the core, which creates a dynamo effect. For reasons that are imperfectly understood, the magnetic field reverses direction every few million years, the magnetic north and south poles switching positions. We have a record of the Earth's magnetic field because of the mineral magnetite, a form of iron oxide. Magnetite is a ferromagnetic material and when it solidifies out of magma its grains become magnetized by the Earth's field, thus recording the magnetic field direction at the time. When scientists started to study the magnetization of the rock close to the mid-Atlantic Ridge, they found a surprising result. Close to the ridge, the magnetite revealed the same field direction as the Earth's field today, but when the scientists looked further away from the ridge the field direction reversed, and reversed again even further from the ridge, producing a zebra-like pattern in the data (Figure 3.5).

By the end of the 1950s, geologists realized the only way to explain the youth of the ocean floor and the strange phenomenon of 'magnetic striping' was if ocean

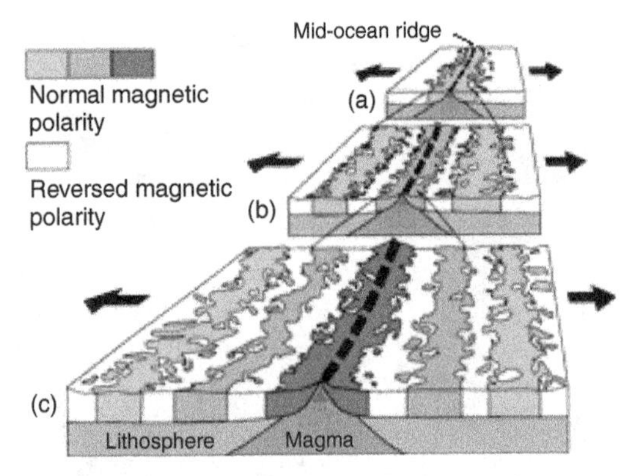

Figure 3.5 Magnetic striping near a mid-ocean ridge (reproduced courtesy of US Geological Survey).

crust is created from magma that wells up at the mid-ocean ridge and, once formed, slowly moves away from the ridge (Figure 3.5). This idea of *sea floor spreading* nicely explains the results. The crust closest to the ridge will be youngest, and thus the field direction revealed by the magnetite is the same as that of the Earth's field today. As one moves away from the ridge, the crust becomes progressively older, and so if one moves far enough away the direction of the field reverses, and reverses again at larger distances. The youth of the crust explains the lack of sedimentary rock and has now been confirmed by radioactive dating (Chapter 7) of rock from the ocean floor.

If the Earth's crust is created at the mid-ocean ridges it must also be destroyed somewhere – otherwise the Earth would be expanding. The theory of *plate tectonics*, which was created in the early 1960s, resembles Wegener's old idea of continental

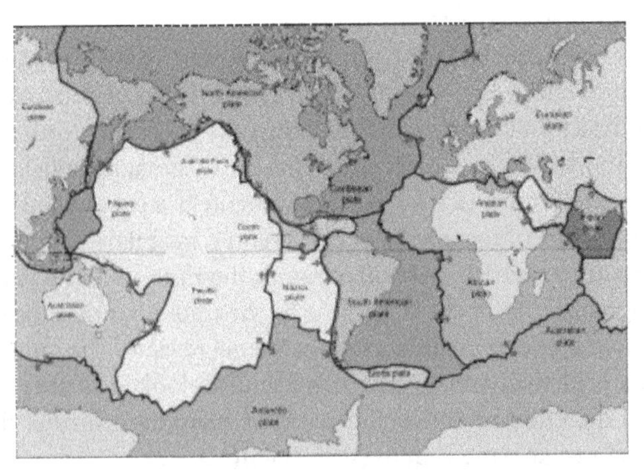

Figure 3.6 The plates forming the Earth's lithosphere (reproduced courtesy of US Geological Survey).

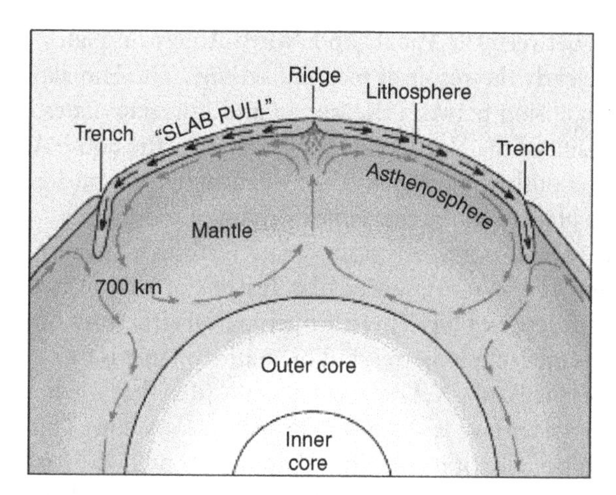

Figure 3.7 A sketch showing the basic ideas of plate tectonics. The arrows show the pattern of convection currents in the asthenosphere (reproduced courtesy of US Geological Survey).

drift, but in plate tectonics it is not the continents that move but plates. According to the theory, the Earth's lithosphere is divided into about 12 plates (Figure 3.6). Rock is added onto the plates at the mid-ocean ridges by magma that is flowing up from the asthenosphere. Plates are destroyed when one plate is forced down under another plate, which occurs at oceanic trenches, deep canyons found at the edges of some oceans (Figure 3.7). The theory answers the question that Wegener couldn't answer. As I showed in Chapter 1, the rock in the Earth's asthenosphere acts like a fluid, slowly moving in response to pressure gradients, and in any fluid convection currents may arise. The explanation of convection is very simple, the example that is always used being the everyday event of boiling some water in a pan. The water immediately above the flame at the bottom of the pan becomes hot, and because hot water has a lower density than cold water, it rises to the top of the pan; the denser cold water sinks down the sides of the pan, starting the water circulating and transporting energy from the bottom to the top of the pan. Convection currents also occur in the asthenosphere because of the heat from the core. As shown in Figure 3.7, the engine that is the most probable explanation of the plate motion is convection currents in the asthenosphere, on which the plates ride. The plate motion is not exactly speedy, typically about 2 cm each year, but over millions of years it is sufficient to move the continents thousands of miles – and thus to explain the pattern of fossils noticed by Wegener.

The theory explains many of the Earth's geological structures and events. Most earthquakes and volcanoes occur along plate boundaries, which is not too surprising when one considers the stresses set up by two huge rigid plates of rock sliding past each other. The famous San Andreas Fault, for example, which passes through San Francisco and which is the continual source of earthquakes, both small and large, is the boundary between the Pacific and North American plates. Many mountain chains are also clearly the result of tectonic activity. The Himalayas are the debris of the head-on collision between the Indian and Eurasian plates (Figure 3.6). The Andes are the result of the Nazca plate slipping below the South American plate; as the Nazca plate is pushed down into the asthenosphere, it pushes upwards on the South American plate, creating this famous chain of mountains.

I have spent several pages describing plate tectonics, because this planet-wide geological system is the most spectacular feature of our planet, apart from the oceans and the presence of life. There are geological structures on Mars and Venus that stretch thousands of kilometres, but on neither planet is there a system of plates similar to that on the Earth. Nevertheless, it is possible that when we look at planets outside the solar system we will see similar tectonic systems. This system of plates is also important because of its role in reshaping our planet. Apart from its roles in shuffling around the continents and in the global carbon cycle (see above), which may have a role in making the Earth a habitable planet (Chapter 9), it is possible that if this tectonic[1] system did not exist there would be no continents at all.

There is an important distinction between the parts of the plates that form continents and the parts that form oceans. One might expect that the detailed gravitational field of a planet would bear some relation to its topography; the gravitational acceleration should be slightly higher over continents and especially over mountain ranges because there is more mass in these places. On Venus, measurements by Magellan showed the expected relationship between topography and gravitational acceleration. On the Earth, surprisingly, this is not the case. Even in the eighteenth century it was already known that the Earth's gravitational field did not increase over mountain ranges. The absence of this relationship is explained by an idea proposed 2500 years ago. The Greek philosopher Archimedes, in the Eureka moment in which he jumped out of his bath tub and ran naked down the street, realized that the mass of a floating object is equal to the mass of the displaced liquid. Let us suppose that a plate is floating in the asthenosphere (Figure 3.8). The volumes of the plate below and above the level of the asthenosphere are V_b and V_a and the mass of the plate is therefore $(V_a + V_b)\rho_p$, ρ_p being its density. According to Archimedes' Principle, this must be equal to the mass of the displaced material from the asthenosphere,

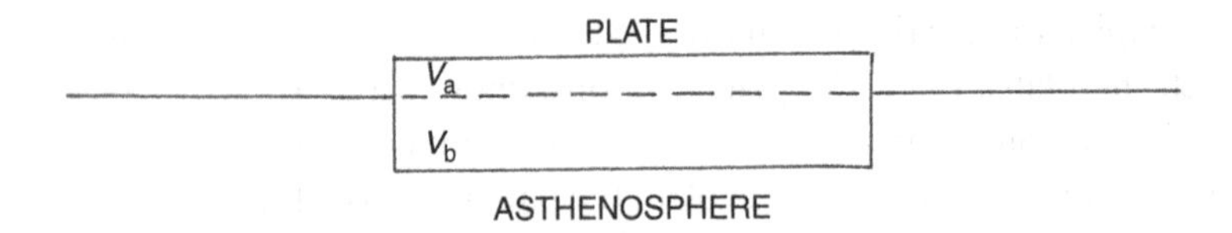

Figure 3.8 A tectonic plate floating in the asthenosphere.

[1]Tectonic comes from the Greek for building, which is apposite given the role of the plate system in building the planet we see today.

which is $V_b\rho_a$, ρ_a being the density of the asthenosphere. Equating the two and rearranging the equation, we obtain an expression for the ratio of the volumes of the plate above and below the level of the asthenosphere:

$$\frac{V_a}{V_b} = \frac{\rho_a}{\rho_p} - 1 \tag{3.1}$$

The equation shows the fraction of the plate above the level of the asthenosphere depends on the density of the plate. It nicely explains why continents are continents. The upper layer of a continental plate is composed of a mixture of sedimentary and granite rocks, which have a lower density than the basaltic rocks that form an oceanic plate; the continents stick up above the surface of the oceans because they are made of lower density rock. Archimedes Principle, which in geophysics is named the *principle of isostasy*, also shows why the gravitational acceleration is not higher over continents; the gravitational acceleration is independent of the type of plate, because the mass of the displaced material is the same as the mass of the plate. The existence of a relationship between topography and gravitational acceleration on Venus is additional evidence that the lithosphere of Venus is not divided into separate tectonic plates floating in an asthenosphere.

The different densities of continental and oceanic rock explains why the floor of the ocean is young and why the oldest rock on Earth is found in the centres of continents. When an oceanic plate (or the oceanic part of a plate which comprises both ocean and continent) collides with a continental plate, it will always be the heavier oceanic plate that sinks down into the asthenosphere (Figure 3.9). This

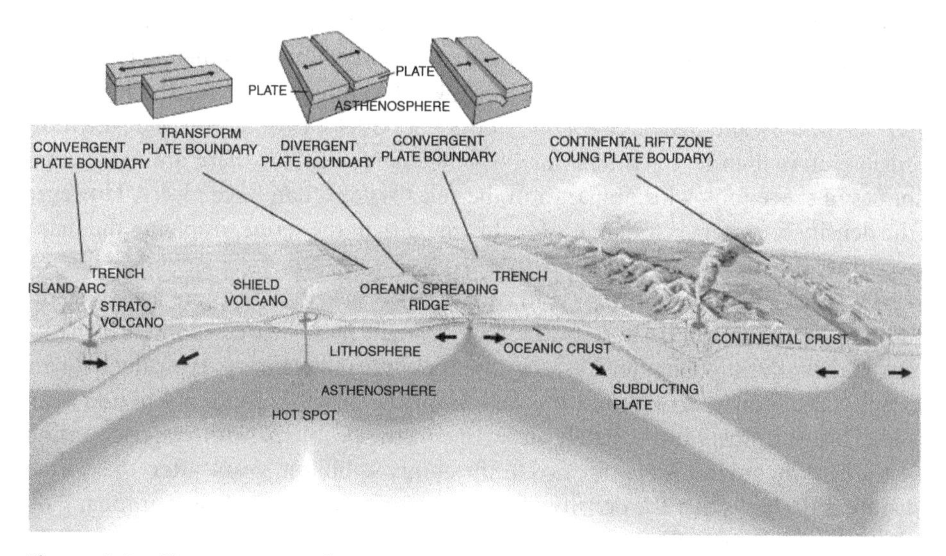

Figure 3.9 The encounter of an oceanic and a continental plate (reproduced courtesy of US Geological Survey).

explains why radioactive dating shows that the ocean floor is never more than 200 million years old, whereas the oldest rock on continents is 3.7 billion years old. Their large ages mean that any continental plate may have grown substantially in thickness since its formation, partly because of the increasing deposits of sedimentary rock but also because of material accumulated from oceanic plates that are being destroyed in the asthenosphere; when an oceanic plate is melted the light magma floats up and sticks to the bottom of the continental plate (Figure 3.9). This gradual growth means that at some time in the past there may have been no continents at all, merely a universal ocean covering the Earth. If, by chance, some oceanic plate managed to endure longer than the others, it might have accumulated enough sediments and low-density igneous rock from the destruction of other plates to remove the possibility of it being destroyed. Eventually, this oceanic plate would have poked its head above the waters and become a continent.

3.3 Crater counting

Let us now consider the surfaces of the other planets. Since it is not possible, at least for the moment, for a human geologist to stroll over the surface and handle the rocks, it is important to understand the limitations of the various methods we have for investigating their surfaces. We will start with the method that has probably kept more planetary scientists busy for longer than any other activity: crater counting.

The basic idea of crater counting is very simple. Because of the continual rain of debris from space, there are more craters on the older parts of a planetary surface, and so it should be possible to estimate the surface's age from the density of craters. Historically this method has been very important because for many years the only information about most of the planets and moons were images of their surfaces (still the case for some objects). Figure 3.10 shows the crater counts for the maria and terrae, the dark and light areas of the Moon. For both types of terrain there are more small craters than large craters, for the obvious reason that there are more small pieces of space junk whizzing around the solar system than large pieces. However, the density of craters is much less for the maria than the terrae, implying the maria were formed after the terrae.

The trick in crater counting is to turn relative ages into absolute ages. We can do this for the Moon because of the rock brought back by the Apollo astronauts. Radioactive dating (Chapter 7) shows the rock from the maria is typically between 3 and 3.5 billion years old and the rock from the terrae is about 4.4 billion years old. These numbers immediately show one of the problems with crater counting. Although the maria were formed only about 1 billion years after the terrae, Figure 3.10 shows that the density of large craters is over 10 times less, implying the impact rate had already declined by a large factor since the formation of the solar system. The uncertainty in exactly how the impact rate has changed with time is why it is not easy to turn relative ages for a surface into absolute ages – unless one

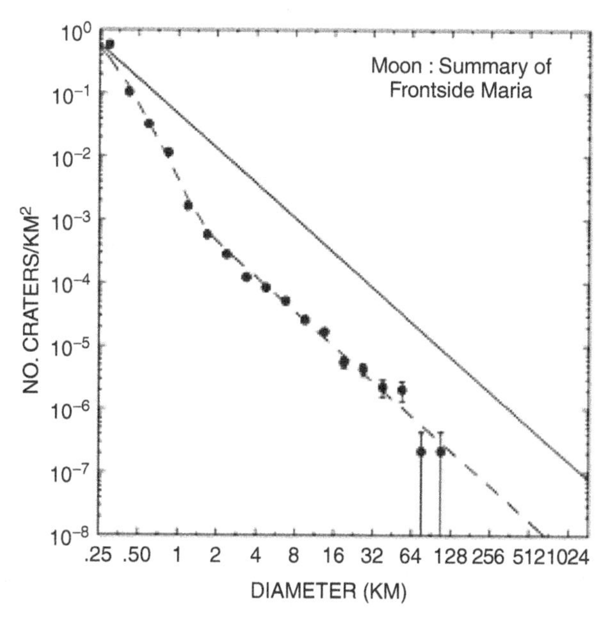

Figure 3.10 The number of craters per square kilometre of the Moon's surface plotted against the diameter of the crater. The straight line shows the counts for the terrae, the dots for the maria.

has been able to bring back rock from the planet and analyse it in the laboratory. Analysis of the moon rock has shown that the maria are largely made up of a dense rock, *mare basalt*, whereas the terrae are made up of *anorthosite*, a rock with a lower density. The maria also have lower elevations than the terrae. One plausible story that explains all these differences is that the Moon started out as a ball of molten rock, the terrae being the first crust that formed as the Moon cooled. The terrae are naturally made out of rock with a lower density because this would have floated to the surface of the ball of magma. Under the surface there would still have been molten rock, and, according to this story, later large impacts punctured holes in the crust, out of which magma flowed to form the maria. These impacts would have been more likely to puncture thin parts of the crust, which is why the maria have lower elevations than the terrae. The lunar terrae are now so covered in craters that any new impacts cover up as many existing craters as they form new ones, and so the crater density for the terrae is the maximum one can get on a planetary surface.

Planetary scientists have made similar analyses of the crater densities on other objects. Figure 3.11 shows the results for the Earth and Mars. The explanation for the low crater density on the Earth is a mixture of erosion, plate tectonics and concealment by the deposit of sediments. In contrast to the Moon, there are as many big craters on Earth as small ones, explained by the fact that it is harder to hide large craters. In the case of Mars, planetary scientists have used a mixture of crater

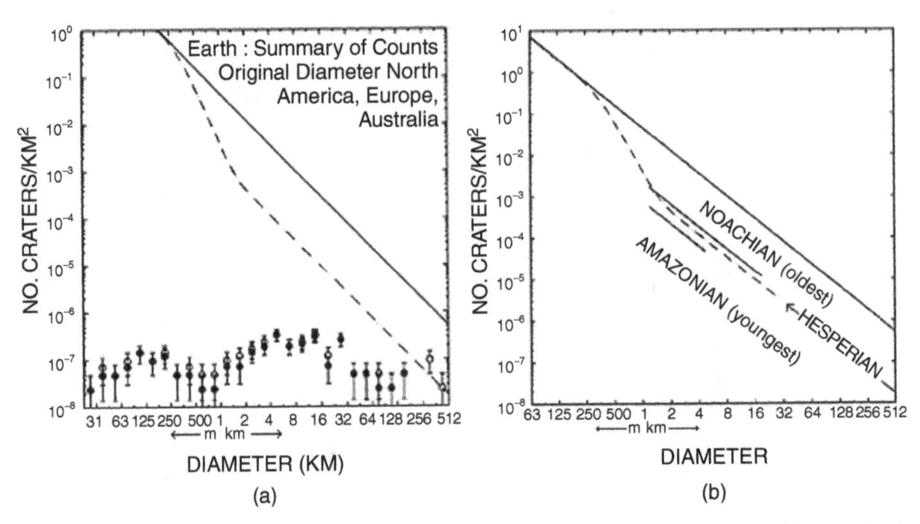

Figure 3.11 The results of crater counting for the Earth (a) and Mars (b). For both figures, the dashed and solid lines are reference lines from the lunar crater counts, for the maria and terrae respectively. The short lines in (b) shows how planetary scientists have used the crater counts to divide the Martian surface into three geological epochs.

counting and other arguments to piece together a geological history for the planet. These other arguments are also based on the images of the planet and include the common sense one that a geological feature that stops unexpectedly must be being concealed by a younger part of the surface. Both the crater counts and these other arguments suggest that the Martian surface was formed in three distinct epochs: the Noachian epoch (4.3–3.5 billion years ago), the Hesperian epoch (3.5–1.8 billion years ago) and the Amazonian epoch (1.8 billion years ago until the present). These ages are very uncertain for two reasons. First, we do not have the independent ages from radioactive dating that we do for the Moon. Second, although it may seem reasonable to assume that a Martian and lunar region with the same crater density must have roughly the same age, this is only approximately true because the amount of debris at the orbit of Mars was probably rather different from that at the orbit of the Earth–Moon system. Figure 3.12 shows a geological map of Mars, based not on the dating of layers of sediments as on the Earth but on the crater counts and on years of careful study of images of the planet.

3.4 Mercury and Venus

I will spend most of the rest of this chapter describing the surface of Mars, about which we know much more today than only a couple of years ago. However, I will briefly describe the surfaces of Mercury and Venus. We will pass over Mercury very

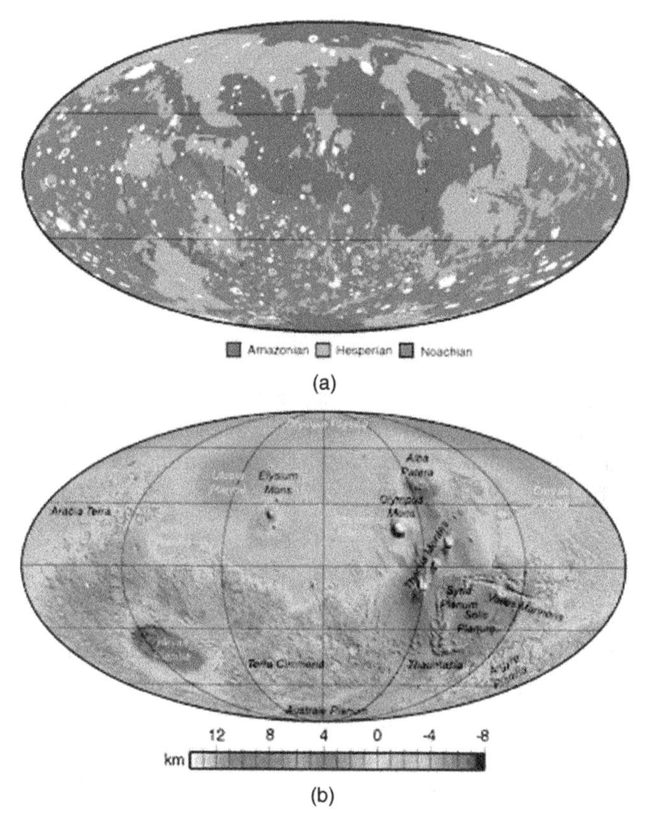

Figure 3.12 Two maps of Mars. (a) is a geological map showing the parts of the surface that were formed during the three epochs of Martian history: Noachian (orange), Hesperian (green), Amazonian (blue). The white areas are where debris from recent large impacts has covered geological structures and earlier craters, making it impossible to estimate the age of the surface beneath. (b) is a topographic map made by the Mars Orbiter Laser Altimeter (see text) (from Solomon *et al.* 2005, *Science*, **307**, 1214 reprinted with permission from AAAS). A colour reproduction of this figure can be seen in the colour section, located towards the centre of the book.

quickly. The crater density measured from the images of the planet (Figure 1.1) shows that the surface of Mercury is probably as old as the oldest parts of the Moon. However, there are no features on Mercury that look like the maria on the Moon. We will know much more about this planet when, after a gap of four decades, it is visited again by Messenger (NASA) in 2011 and by BepiColombo (ESA) in 2019.

Most of what we know about the surface of Venus comes from Magellan, which in the early 1990s used radar to map the surface of the planet with a resolution of about 100 km. We also know a little about the rocks from the Russian Venera spacecraft that have landed on the planet – these appear to resemble terrestrial

Figure 3.13 Topographic map of Venus made by Magellan, in a mercator projection (north is at the top). The lowest regions are shown as blue, the highest as red (courtesy: NASA). A colour reproduction of this figure can be seen in the colour section, located towards the centre of the book.

basalts. Figure 3.13 shows a topographic map of the planet made by Magellan. The highest region on the planet is Maxwell Montes (the Maxwell Mountains) at the top of the map. The difference in height between this region and the lowest point on Venus is about 13 km, which is similar to the Earth – not surprising given the similar sizes of the two planets and the argument about the characteristic size of the mountains on different planets that we considered in Chapter 1. The density of craters, which is fairly similar all over the planet, implies that all parts of the surface have a similar age, somewhere between a few hundred million years and 1 billion years. There are many interesting geological features (e.g. Figure 1.2) that do not look much like anything seen on the Earth, but which planetary scientists, for want of anything better, are forced to try to explain by extrapolation from geological processes that do occur on the Earth. A popular explanation of the pancake domes shown in Figure 1.2, for example, is that these are places where magma flowing up from the interior has distended the surface, which has then collapsed when the magma has retreated. There does not appear to be a planet-wide geological structure like the system of tectonic plates seen on the Earth. A related difference between the two planets is that on Venus the gravitational acceleration *does* increase in the mountains, showing that the lithosphere is not divided into plates floating in the asthenosphere. As the masses of the two planets are very similar, the thickness of the lithosphere should also be similar (Chapter 1), and it is not obvious at the moment why on one planet the lithosphere should be divided into plates and not on the other. One possible explanation of the relative youth of the surface (compared to Mercury and most of Mars) is if stresses within the lithosphere occasionally cause it to rupture completely; the lithosphere suddenly (in geological terms) breaks into bits, which then sink into the asthenosphere, to be replaced by a new lithosphere made from magma flowing up from the interior. If true, there *was* tectonic activity on Venus in the past, which suddenly resurfaced the planet, in contrast to the continuous resurfacing of our planet being carried out by the Earth's tectonic system.

3.5 A tourist's guide to Mars

We will now consider the most interesting of the planets, Mars, and especially the discoveries made about the planet during the last 2 years. More spacecraft have been sent to Mars than any other planet (Table 3.2, which also illustrates the heartbreak of planetary exploration as several of the missions were failures), for the obvious reason that it is the planet most similar to the Earth. It is, admittedly, not *that* similar to the Earth – it is much colder and has a much thinner atmosphere composed mostly of carbon dioxide rather than oxygen – but it is at least less obviously hostile to life than killer planets like Venus. There is also the promise of water, a necessity for life, at least as we know it on Earth. The polar caps of Mars contain water, and images taken by the Viking spacecraft in the 1970s showed many features that looked as if they had been produced by water running over the surface. There are more of these features found in the oldest (Noachian) parts of the surface, implying that early in the history of Mars there may have been rivers and even oceans. The atmosphere of Mars must then have been much thicker – otherwise this water would rapidly have evaporated – and so it is likely that because of the greenhouse gases in the atmosphere the planet was much warmer than it is today. It is possible that some of the more unusual organisms on Earth (some of the extremophile bacteria, for example – Chapter 9) could survive on Mars today, but there are definitely ones that would have been able to survive during this earlier epoch. If this early 'warm and wet' period actually happened – and until very recently this has only been a hypothesis based on the geological features seen in the Viking images – there are two obvious big questions: (i) Where is all the water now? (ii) What happened to the atmosphere?

Before I describe the recent exciting discoveries about Mars, we will anticipate the interplanetary tours that may be occurring in a 100 years time and take a quick tour of the planet. Figure 3.14 shows an image of the planet made by Viking in the 1970s. The top Martian tourist attraction will probably be the Valles Marineris, a system of canyons that stretches roughly 4000 km across the planet and can be seen halfway down the image. This structure dwarfs similar structures on the Earth. It is roughly seven times deeper than the Grand Canyon, and if the Valles Marineris was in the United States it would extend from coast to coast. Although there is not a system of tectonic plates on Mars, the Valles Marineris, which looks like a zip across the planet, shows all the signs of having been caused by some colossal upheaval in the planet's lithosphere. The second stop on a tourist's itinerary would probably be one of the Martian volcanoes. The largest volcano in the solar system, Olympus Mons (Figure 1.3), is just off the image, but two huge shield volcanoes can be seen at the top left (the dark ovals). These are not quite as large in volume

Table 3.2 Recent missions to Mars.

Mission	Date of arrival in martian orbit	Description
Mars Observer (US)	1993	Communication lost just before arrival at Mars
Mars Global Surveyor (USA)	1997	Orbiting spacecraft that surveyed Mars for 9 years until November 2006
Mars Pathfinder (USA)	1997	Lander and six-wheeled rover, named Sojourner
Nozumi (Japanese)		Launched in 1998, it passed 1000 km from Mars but failed to achieve an orbit around the planet
Mars Climate Orbiter (USA)	1998	Designed to study the martian climate and atmosphere, it was destroyed when a navigation error caused it to enter the atmosphere at the wrong altitude
Mars Polar Lander (USA)	1999	Intended to land close to the south pole, it lost communication with the Earth just before entering the atmosphere
Mars Odyssey (USA)	2002	An orbiting spacecraft surveying the surface, it is also the communications hub for the Mars rovers (see below)
Mars Express (Europe)	2004	A double mission: an orbiting spacecraft, which is surveying the surface with enough resolution to see a large truck, and a lander (Beagle), which lost communication with the Earth after entering the atmosphere
Mars Rovers – Spirit and Opportunity (USA)	2004	Robot geologists that landed in January 2004 and are continuing to study the Martian rocks and soil
Mars Reconnaissance Orbiter (USA)	2006	An orbiting spacecraft designed to survey the surface with enough resolution to see objects 1 m in size (about the size of a Martian?)
Phoenix Mars Mission (USA)	2008	A mission designed to investigate the water content and look for complex organic compounds in the soil in the northern arctic plains of Mars

Figure 3.14 Image of Mars made by Viking (courtesy: NASA). A colour reproduction of this figure can be seen in the colour section, located towards the centre of the book.

and area as Olympus Mons, but because they are on a swelling in the Martian crust, their summits are almost at the same height.

The most puzzling geological feature on the planet, however, would not be on any tourist's itinerary because it is not visually spectacular, and indeed it is not visible at all in the image of Mars in Figure 3.14. It shows up extremely well, however, in a map made by the Mars Global Surveyor. One of the instruments on this spacecraft was the Mars Orbiter Laser Altimeter (MOLA), which was designed to map the height of the Martian surface, which it did by bouncing infrared laser pulses off the surface and measuring the time they took to return to the spacecraft. Figure 3.12b shows the topographic map of the planet made with MOLA. The map shows very nicely the Valles Marineris and the volcanoes, but it also reveals the intriguing fact that there is a systematic difference in height between the northern and southern hemispheres of Mars: on average, the northern hemisphere is about 6 km lower than the southern hemisphere. And so there is another big question about Mars: what caused this systematic difference between the two hemispheres? Since the geological map in the same figure implies that the surface of much of this low-lying area was formed in the most recent (Amazonian) geological epoch, one might conclude that the difference between the two hemispheres must have been created by some geological event within the last 2 billion years. I will show below that a very recent discovery suggests this is not actually true. Some scientists have suggested that this large low-lying northern area, the Vastitas Borealis, may have been the location of an ocean during Mars' early wet and warm period, and they claim to have found features in images that represent the shoreline of this ancient ocean.

3.6 Recent research on Mars

The new discoveries about Mars have been made with a variety of new instruments that have taken the study of planets well beyond the era of crater counting. Some of the most important are still cameras, but they are spectacularly improved compared to the cameras on previous spacecraft. At the Martian surface, the resolution of the cameras on Mars Express, for example, which has been surveying Mars since 2004, is about 15 m, which means it would be possible for Mars Express to pick out a large truck on the surface. The cameras on the Mars Reconnaissance Orbiter, which arrived in orbit around Mars in November 2006, can see features on the Martian surface as small as 1 m in size, enough resolution to see a Martian, if one exists. Mars Express has two cameras, which like our eyes effectively gives it depth perception and the ability to take images that show the surface in three dimensions.

Figure 3.15 shows several images from Mars Express, all of which show that water is either there now or was once present. Figure 3.15a shows some channels and deposits that are impossible to explain without the sudden flow of a large volume of water across the surface. Figure 3.15b shows a deposit of ice in a crater in the Vastitas Borealis. Figure 3.15c shows a mixture of dust and ice close to the north pole of Mars; Figure 3.15d shows a region close to the equator which scientists suspect may be a frozen sea covered by dust. All of these images agree with the 'wet and warm' theory, in which water once flowed across the surface of Mars but is now all locked up in ice.

Improved cameras have greatly enhanced our knowledge of Mars, producing a flood of data for planetary scientists, but it is new observing techniques that have led to several breakthroughs in our understanding of the planet. Both Mars Express and Mars Reconnaissance Orbiter carry low-frequency radar experiments that allow them to probe several kilometres below the surface of the planet. The spacecraft transmit low-frequency radio waves towards the surface; some fraction of them are reflected back by the surface, while the rest penetrate the surface and are reflected from underground structures. By analysing the times at which the radio waves arrive back at the spacecraft, scientists can investigate the planet's interior down to a depth of a few kilometres. The ESA scientists have used the radar experiment on Mars Express to make two big discoveries. First, they have discovered that the crust in the northern hemisphere is not as young as previously believed. The low density of craters in the Vastitas Borealis had led scientists to conclude that the surface there was mostly formed in the Amazonian epoch. However, the recent radar measurements have shown that there are a large number of craters hidden under the surface in the northern hemisphere, and once these are taken into account, the crater densities in the south and the north are similar. Therefore, although the surface in the north *is* young, there is an older surface hiding beneath. The implication of this discovery is that the difference in height between the northern and southern hemispheres must have been established early in the history of the planet. The second big discovery is that the ESA scientists have measured the thickness of the southern polar cap. It has

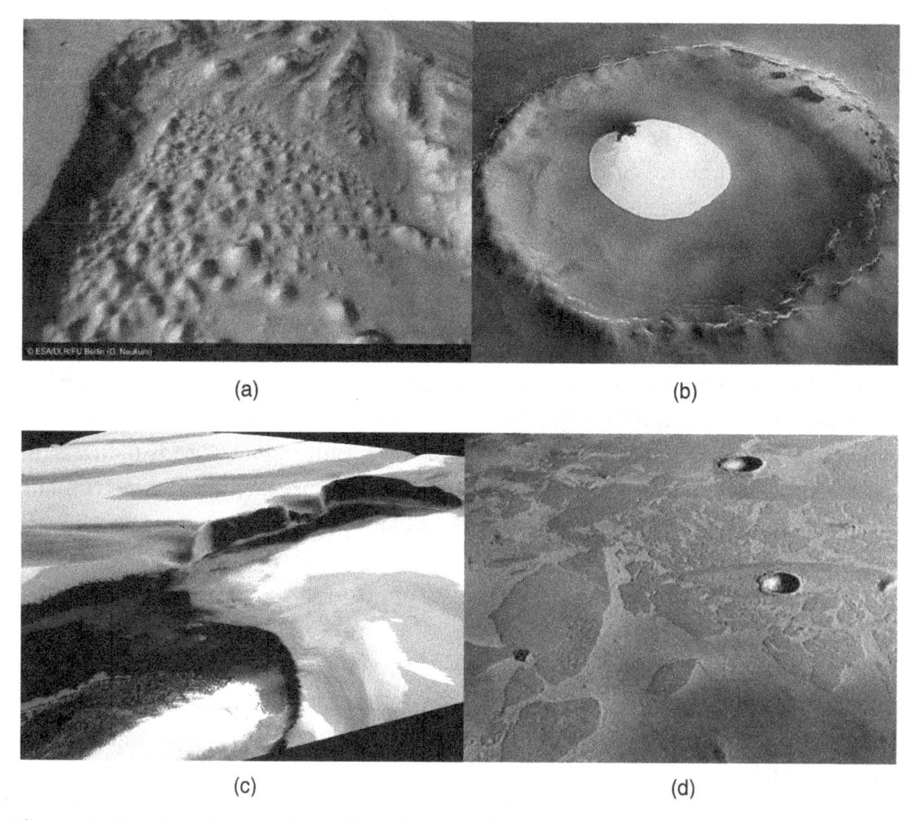

Figure 3.15 Four images from Mars Express that show the presence of water on Mars: (a) an outflow channel; (b) ice in a crater; (c) a mixture of ice and dust at the North Pole; (d) a possible frozen sea covered by dust (courtesy: ESA).

always been known that the Martian polar caps contain a mixture of frozen water, frozen carbon dioxide and dust (Figure 3.15), but it has never been clear how much water there is. The ESA scientists have used the Mars Express radar to show that the layer of ice is almost 4 km thick and that 90 % of it is frozen water. The amount of ice is equivalent to a layer of water 11 m thick covering the entire planet; if placed in the Vastitas Borealis, the water would make a respectable ocean.

The biggest recent discoveries, however, have come from the mineralogical surveys. Without 'sample return' – actually bringing the rock back to the Earth and testing it in a laboratory – the only way scientists can determine the composition of the rock on a planet is to use spectroscopy. The orbiting spacecraft, Mars Express and Mars Reconnaissance Orbiter, and NASA's two robot geologists, Spirit and Opportunity, have all used similar instruments to investigate the composition of the Martian rock. The advantage that Spirit and Opportunity have of actually being on the planet is that they can handle the rock; they both have a rock abrasion tool which allows them to remove part of a rock's surface, which may have been physically or chemically altered by the atmosphere, and which is exactly what a human geologist would do. They can also obviously study the rock in much more detail than an orbiting spacecraft. Their disadvantage is that they can only tell us about the rock within a few kilometres of their landing sites, whereas the orbiting spacecraft can survey the whole planetary surface. A third robot geologist, although this time a stationary one, landed on Mars in May 2008. The aim of the Phoenix Mars Mission is to study the soil in the northern arctic regions, in particular its water content and whether it contains complex organic compounds.

There are two very different types of spectroscopy used by planetary scientists. Standard astronomical spectroscopy in the optical and infrared wavebands, in which one looks for the spectral lines produced by different compounds, is a powerful way of investigating the minerals that make up the rock, although its disadvantage is that it is only possible to detect minerals with spectral features within the instrument's wavelength range. An alternative is to use a gamma-ray or neutron spectrometer to determine the elements, rather than the compounds, that make up the rock. These instruments detect the gamma rays or neutrons emitted by atomic nuclei on the surface that have been excited by the cosmic rays that bombard the surface of every planet; since each element produces gamma rays and neutrons with different energies, by measuring their energies one can determine the elements making up the surface rock. The disadvantage of this type of spectroscopy is that merely knowing the elemental composition of a lump of rock does not necessarily tell you the mineralogical composition – different minerals may contain the same mixture of elements. One place where neutron spectrometry has proved very effective is the Moon. The Lunar Prospector spacecraft detected a large amount of hydrogen at the lunar poles. In this case, the only plausible chemical compound containing hydrogen is water, and these measurements imply there is between 10 and 300 million tonnes of ice on the Moon.

For investigating the surface of Mars the most useful type of spectroscopy has been the standard astronomical type. Figure 3.16 shows two maps of the surface made with one of the spectrometers on Mars Express. Figure 3.16a shows the distribution of the mineral pyroxene, an important constituent of basaltic rocks. If you compare this with the geological map of Mars shown in Figure 3.12, you will see that this mineral is mostly found in the older (Noachian and Hesperian) parts of the surface. Both pyroxene and olivine, another important constituent of basaltic rock, have been detected in these regions, implying the older parts of the surface are mostly basalt. Figure 3.16b shows the distribution of iron oxides, the compounds that give Mars its red colour (iron oxide is just the posh name for rust). These are found mostly in the younger (Amazonian) part of the surface. There is also no sign from the spectroscopy of any hydrated minerals, minerals that incorporate water, in these younger regions. According to the crater counts, the Amazonian epoch started

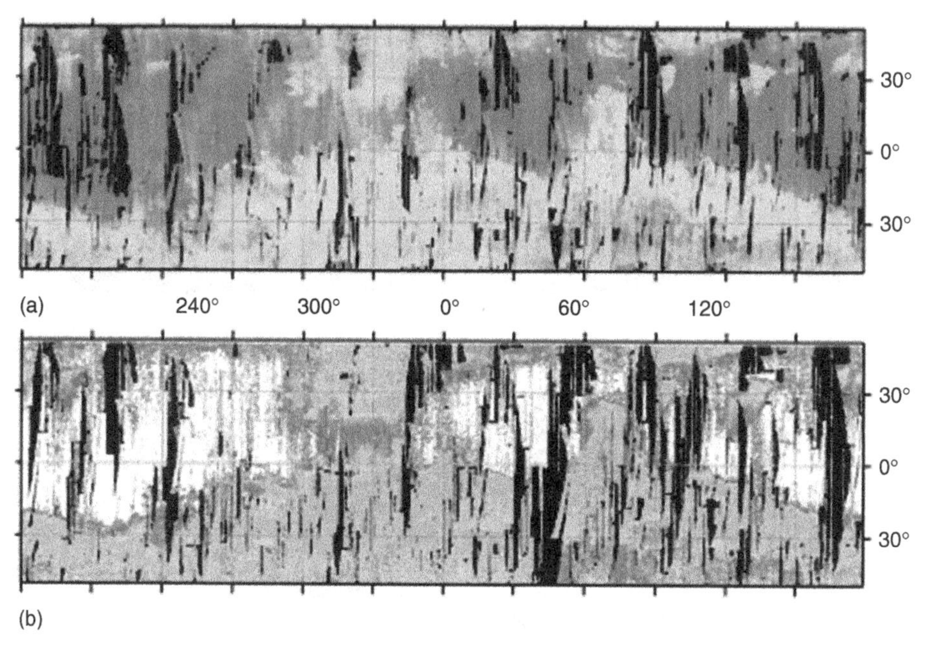

Figure 3.16 Maps made by Mars Express of the distribution of two important minerals (Mercator projection, north at the top). (a) shows the distribution of pyroxene, an important constituent of basalt rocks, with yellow showing its presence and blue where there is none. (b) shows the distribution of iron oxide, with white and red indicating its presence (from Bibring *et al.* 2006, *Science*, **312**, 400 reprinted with permission from AAAS). A colour reproduction of this figure can be seen in the colour section, located towards the centre of the book.

about 1.8 billion years ago, and therefore this result from Mars Express implies that for the last 2 billion years Mars has been very dry.

The most exciting recent discoveries, however, have been of new evidence that water did once flow over the Martian surface. Figure 3.17 shows an image taken by one of the NASA rovers of Burns Cliff, a rock outcrop that is part of the rim of Endurance Crater. The layers visible in the image look remarkably like the layers of sedimentary rock that one sees in cliff faces on Earth. The instruments on the rover showed that the rock is sandstone and that the grains in the rock are rich in sulfates, salts that could only have been formed in the presence of water. The rock face also contains large numbers of 'blueberries', round objects between 4 and 6 mm in diameter that are rich in the mineral haematite; haematite is another mineral that (on Earth at least) is always formed in the presence of water. One of the best ways to decipher the geology of a planet is to walk over the surface and handle its rock, and the Mars rovers have done this almost as well as a human geologist. Both rovers have now found many signs that the surface of the planet was once wet.

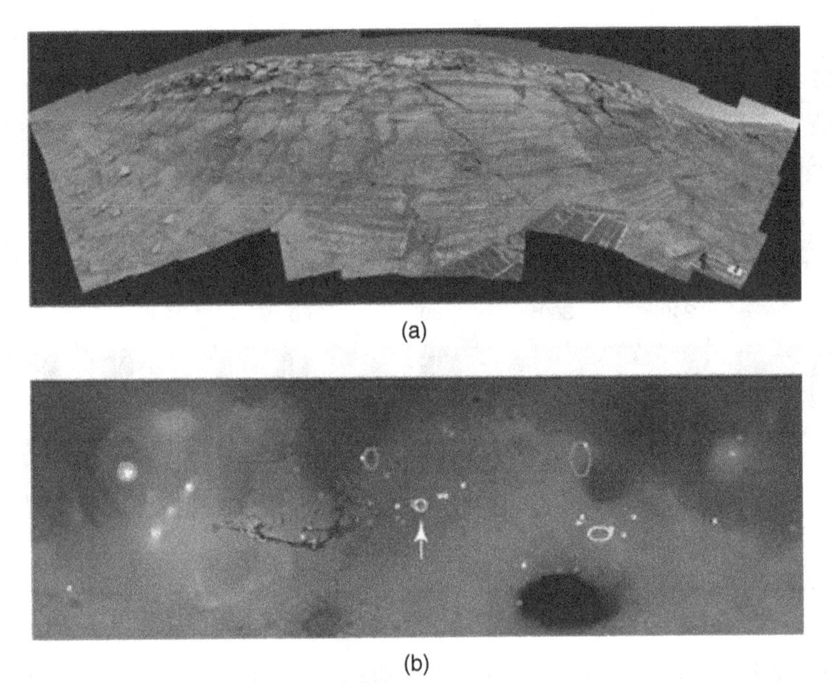

(a)

(b)

Figure 3.17 A local and global view of the geology of Mars. The local view in (a) is an image taken by the Mars rover Opportunity of a few metres of Burns Cliff, a rock outcrop that is part of the rim of Endurance Crater. The global view in (b) shows a topographic map (lighter colours imply a higher elevation) on which are superimposed the location of several important minerals: sulfates (blue), phyllosilicates (red), other hydrated minerals (yellow). The arrow shows where Opportunity landed (from Bibring *et al.* 2006, *Science*, **312**, 400 reprinted with permission from AAAS). A colour reproduction of this figure can be seen in the colour section, located towards the centre of the book.

The Mars rovers have only been able to study a few kilometres of the surface. The instruments on Mars Express have given us a panoramic view of the geology of the entire planet. Figure 3.17b shows the location of various minerals revealed by the spectroscopy. These locations are superimposed on a topographic map of the planet (the north–south asymmetry and the positions of the large volcanoes can be seen clearly). The red lines show the presence of phyllosilicates, the blue lines the presence of sulfates and the yellow lines other hydrated minerals. The discovery of these minerals implies there was once water on the Martian surface. Phyllosilicates, for example, are simply clays, which on Earth are formed by the deposition of sediments in standing bodies of water. Their presence on Mars implies that there were once lakes, oceans or rivers on the planet.

The location of these minerals has filled in some of the details of the history of Mars, and in particular the history of its water. None of these minerals has been found on the younger surface to the north, which confirms the early warm and wet theory – there were once lakes, rivers and oceans on Mars but this was billions of years in the past. The phyllosilicates are only found in the oldest terrain of all, the Noachian surface, which, according to the crater counts, was formed between 4.3 and 3.5 billion years ago. The sulfates are mostly found on the younger (3.5–1.8 billion years) Hesperian surface. The Mars Express team has suggested one possible explanation for this difference. The formation of sulfates requires sulfur and acidic conditions, both of which are produced by volcanic activity. The required conditions might have been produced if the large volcanoes on Mars were formed at the end of the Noachian period.

I have spent many pages describing the spectacular recent discoveries on Mars, because they are a good illustration of the power of the new techniques and instruments that have been developed for observing the planets, and Mars is the first planet for which the full range of these techniques has been used. You may have noticed that there are many big questions for which we still do not have answers. We do not know, for example, how two of the most spectacular geological structures on Mars, the north–south asymmetry and the Valles Marineris, were formed. We also cannot answer a fundamental question about the atmosphere. The new results are additional evidence that there was an early wet and warm period on Mars, implying the existence then of a much denser atmosphere. But where has this atmosphere gone? On the Earth, most of the carbon dioxide is locked up in limestone and chalk (see above), but this does not appear to be the case on Mars because carbonate rocks produce a distinctive spectral feature that should be easy to detect with Mars Express, and none has been seen. An alternative hypothesis is that the atmosphere was gradually lost because of Mars' relatively weak gravitational field. At the moment, there is evidence, which I will describe in Chapter 5, both for and against this hypothesis. Another question that has only been partially answered by Mars Express is how much water there is on Mars. We now know fairly accurately the amount of water locked up in the southern polar cap, and there are indications of ice elsewhere on the planet, but we still do not know exactly how much water there is. The biggest questions of all are about life. It seems likely that Mars would have been a hospitable place for life during the Noachian epoch. Did life actually start then, and if so is there still life somewhere on Mars? A good place to start looking for an answer would be places where Mars Express has shown there are phyllosilicates, because even if there is no life on Mars today there might be signs there that there was life in the past. During the next few decades, as the exploration of Mars intensifies in preparation for a human landing, some or all of these questions should be answered.

Exercises

1 If the typical speed of plates is about 5 cm per year, estimate the typical time between the creation of a plate and its destruction. The oldest rock on Earth is in Canada and is about 3.7 billion years old. Explain why this is much larger than the time you estimated in the first part of the question (radius of Earth: 6.4×10^6 m).

2 The upper part of the lithosphere consists of the Earth's crust, which consists of the distinctive rocks that make up the ocean floor and the continents. The thickness of the oceanic crust is about 10 km. The continental crust is typically about 5 km higher than the floor of the ocean. On the assumption that both the continental and oceanic crust are floating directly in the asthenosphere, estimate the thickness of the continental crust. You may assume that the density of the continental crust, oceanic crust and the asthenosphere are 2800, 3000 and 3300 kg m^{-3}, respectively. You should ignore the effect of the oceans on this calculation.

Further Reading and Web Sites

Bibring, J.-P., Langevin Y, Mustard J.F. *et al.* (2006) Global mineralogical and aqueous Mars history derived from OMEGA/Mars Express data. *Science*, **312**, 400.

Mars Express web site http://www.esa.int/esaMI/Mars_Express/index.html. (accessed 17 September 2008)

Mars Reconnaissance Orbiter web site http://mars.jpl.nasa.gov/mro/. (accessed 17 September 2008)

Mars Rover web site http://marsrovers.nasa.gov/home/index.html. (accessed 17 September 2008)

Squyres, S. *et al.* (2006) Two years at Meridani Planum: results from the Opportunity Rover. *Science*, **313**, 1403.

Watters, T.R., Leuschen, C.J., Plaut, J.J. *et al.* (2006) MARSIS radar sounder evidence of buried basins in the northern lowlands of Mars. *Nature*, **444**, 905.

4 The interiors of the planets

Specialisation is for insects
Robert Heinlein

4.1 What we do and don't know about planetary interiors

In this chapter I am going to take off a geologist's hat and put on my physicist's hat again. In the last chapter we considered the variety of interesting geological structures that one sees on the surfaces of the planets, but why are the surfaces of the Earth and the other inner planets such interesting places? Part of the answer must be that the interiors of the planets are hot, which on Earth leads to convection currents in the asthenosphere and the cornucopia of geological structures caused by plate motion, and which is also undoubtedly ultimately responsible for the interesting surfaces of Venus and Mars. However, this cannot be the whole answer because the Sun's interior is also very hot but its photosphere is nowhere near as interesting as the surface of a planet. The geologist's approach does not work here, and as you will see at the end of this chapter, the reason the Earth is such a nice place for a geologist to walk about on is a simple piece of physics.

I also want to consider the less exciting but still important question of how, stuck as we are on the surface of a planet, we can find out about its interior, let alone the interiors of the other planets. The study of methods and their limitations is important for a scientist, because one always needs to evaluate the work of other scientists. When faced with a diagram in a textbook or a scientific paper that purports to show the structure of Jupiter, for example, how seriously should you take it? Is this likely to be the real structure of Jupiter or is this just one of several

Planets and Planetary Systems Stephen Eales

possible models that fit the data, or is the author even trying to peddle his own pet theory, for which there is actually not much evidence at all? In science, as in life, informed scepticism is usually a good thing.

I will delay considering the Earth's interior, about which we know much more than the interiors of any of the other planets because of the seismic stations dotting its surface, until later in the chapter. Seismic data has also revealed a little about the Moon's interior, because five of the Apollo space missions carried seismometers, which continued to operate after the astronauts left until 1977, but we have no seismic information for any other planet or moon. I will start by considering what we can learn about a planet without the aid of seismic data, which is the situation we are in for all the other moons and planets in the Solar System.

I will use as examples the largest planet and moon in the Solar System: Jupiter and Ganymede. Ganymede is one of the four giant moons of Jupiter (there are at least 59 others) discovered by Galileo when he looked at the planet with one of the first telescopes in 1609. Two basic things that we know about both objects are their masses and their volumes, which when combined yield their densities. The average densities of Jupiter and Ganymede are $1326\,\mathrm{kg\,m^{-3}}$ and $1940\,\mathrm{kg\,m^{-3}}$, which immediately tells us something about their interiors because both values are much less than the typical density of rock ($\approx 3000\,\mathrm{kg\,m^{-3}}$). Jupiter's low density is unremarkable because it is a gas giant, but Ganymede appears to be a solid object and so its low density is a little surprising. It must contain a component with a lower density than rock, and given the abundances of elements in the Solar System (Table 1.2), there is only one real possibility: ice. This has a density of $\approx 1000\,\mathrm{kg\,m^{-3}}$, and the low density of Ganymede may be explained by a model in which 60 % of the moon's mass is rock and 40 % is ice.

This is interesting, but it is a long way from a detailed model of the interior of the moon. The value I have used for Ganymede's density was derived from a measurement of its mass made by NASA scientists as they tracked the spacecraft Galileo when it passed close to the moon in June and September 1996. Galileo's trajectory also gave one important piece of extra information about the interior of the moon.

The gravitational potential of a planet or moon satisfies Laplace's equation (for those of you who have never seen an equation like this, don't panic):

$$\nabla^2\Phi = 0 \tag{4.1}$$

The solution of this equation for an axisymmetric object like a planet or a moon is

$$\Phi(r,\theta,\phi) = -\frac{GM}{r}\left[1 - \sum_{2}^{\infty}\left(\frac{R_e}{r}\right)^n J_n P_n(\cos(\theta))\right] \tag{4.2}$$

This looks quite complicated, but it is not as scary as it first looks. The three coordinates – r, θ, ϕ – are the coordinates used in the spherical polar coordinate system, which is the sensible one to use for an object that is approximately spherical such as a planet or moon; M is the mass of the moon; R_e is its equatorial radius; and the terms $P_n(\cos\theta)$ are a set of mathematical functions called *Legendre polynomials*. The remaining terms are the numbers, J_n, by which each Legendre polynomial is multiplied and which express the contribution of that particular Legendre polynomial to the overall gravitational potential.

Most of this you may promptly forget. The equation can be summed up by the statement that any gravitational potential can be expressed as an infinite sum of terms, with the strengths of the terms being represented by a set of coefficients, J_n. For real objects like planets, however, only a few of these terms are ever important. The gravitational potential of a sphere is equal to $-GM/r$, and so if a planet were a perfect sphere, all the terms in Equation 4.2 would be zero apart from the first. Real planets, however, are almost perfect spheres, and the only coefficient that is usually important is the second one, J_2. The NASA scientists were able to use the trajectory of Galileo as it passed by Ganymede to estimate the value of J_2 for the moon.

Their estimate contained valuable information about the moon's interior. The value of J_2 is related to the difference between the polar and equatorial moments of inertia of the moon. The moment of inertia of a system of masses, m_i, is defined as

$$I = \sum_i m_i r_i^2 \tag{4.3}$$

in which r_i is the perpendicular distance from the axis of rotation. This definition shows that the moment of inertia depends on the choice of the axis, and so the moment of inertia of a planet around a north–south axis (I_p) will not necessarily be the same as the moment of inertia around an axis in the equatorial plane (I_e), although they are very similar. Its derivation is beyond the scope of this book, but there is a relation between J_2 and the difference between the two moments of inertia:

$$J_2 = \frac{I_p - I_e}{MR_e^2} \tag{4.4}$$

The reason planets are not perfect spheres is rotation; the effective gravitational force is less at the equator than the poles because of the centrifugal force and so a planet's polar radius is slightly less than its equatorial radius. A useful measure of this effect is the dimensionless parameter, Λ, which is the ratio of the centrifugal force at the equator to the gravitational force:

$$\Lambda = \frac{\omega^2 R_e^3}{GM} \tag{4.5}$$

in which ω is the rotational angular velocity of the planet. There is also a relationship between Λ and the difference in the two moments of inertia, although it is not a simple equation like Equation 4.4. By combining both relationships, it is possible to show there is an approximate relationship between the moment of inertia of a planet or a moon and the values of J_2 and Λ:

$$\frac{I}{MR^2} \simeq \frac{\frac{2}{3}J_2}{J_2 + \frac{1}{3}\Lambda} \tag{4.6}$$

In this equation, I have dropped the suffixes for polar and equatorial, because the polar and equatorial radii and moments of inertia are so similar. The Galileo team used this equation to estimate the moon's moment of inertia:

$$\frac{I}{MR^2} = 0.3105 \pm 0.0028 \tag{4.7}$$

It is easy enough to show (see Exercises) that if the density of Ganymede were everywhere the same, the value of the ratio above would be exactly 0.4. The fact that it is different from this gives us some additional information about the moon's structure. When faced with this kind of situation, the obvious thing to do for any scientist is to make a model, and we will follow in the steps of the NASA scientists and use this information to build a simple model for the interior of Ganymede.

We also have one extra piece of information, also discovered by Galileo (the spacecraft not the scientist): Ganymede has a magnetic field. This suggests that the moon has an iron core. Let us assume that Ganymede is divided into three layers: an iron core, a layer of rock, and a surface layer of ice (Figure 4.1). Let us also assume that the density within each layer is constant, so that the density of the iron core is 8000 kg m^{-3}, the density of the rock layer is 3000 kg m^{-3} and the density of the ice is 1000 kg m^{-3}. At first sight, this may seem inconsistent with the principle of hydrostatic equilibrium, which implies that to balance the gravitational field the pressure in the moon must increase with depth (Chapter 1), but pressure is largely independent of density for a solid, and so it is still possible for the density in a layer

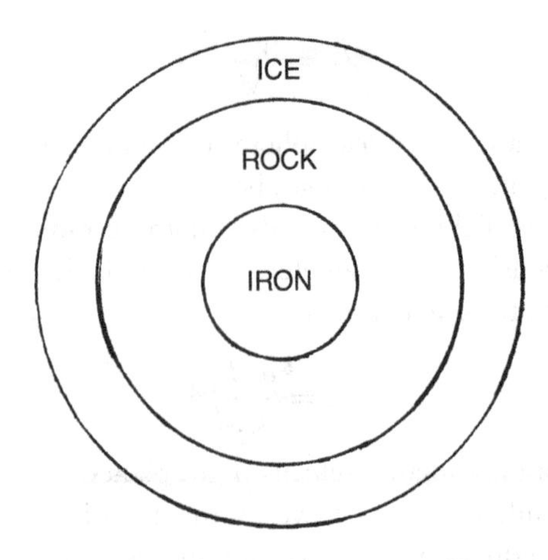

Figure 4.1 A back-of-the-envelope model of the interior of Ganymede.

to be roughly constant and still have the pressure increase with depth in the way necessary to support the weight of the moon.

The two unknowns in the model are the radius of the core, R_c, and the outer radius of the rock mantle, R_m. We can estimate these by requiring that the model reproduces Ganymede's mass and moment of inertia. The mass is the sum of the masses of each layer, which is simply the volume of the layer times its density:

$$M = \frac{4\pi}{3}\left[8000R_c^3 + 3000\left(R_m^3 - R_c^3\right) + 1000\left(R_g^3 - R_m^3\right)\right] \tag{4.8}$$

in which R_g is the radius of the moon. The moment of inertia is also the sum of the moments of inertia of the three layers. The moment of inertia of a thin spherical shell of thickness δr and density ρ is $(8\pi\rho r^4\delta r)/3$, and so the moment of inertia of a thick shell of inner radius R_A and outer radius R_B is (Those without calculus should avert their eyes from the middle part of the equation and only look at the result):

$$I = \int_{R_A}^{R_B} \frac{8\pi r^4 \rho}{3}\,dr = \frac{8\pi\rho}{15}\left(R_B^5 - R_A^5\right) \tag{4.9}$$

The moment of inertia predicted by

$$I = \frac{8\pi}{15}\left[8000R_c^5 + 3000\left(R_m^5 - R_c^5\right) + 1000\left(R_g^5 - R_m^5\right)\right] \tag{4.10}$$

Equations 4.8 and 4.10 are a pair of simultaneous equations with two unknowns, R_c and R_m. It is not possible to solve these with pen and paper, but it is easy enough to solve them by writing a short computer program. The answer is that the core extends about 30 % of the way to the surface, the rock mantle finishes about 75 % of the way to the surface, and the remainder of the moon (a layer 700 km thick) is ice.

However, remember the importance of informed scepticism. The model is based on only three pieces of information: the mass, the moment of inertia and the existence of a magnetic field. Some of the assumptions of the model might easily be wrong. It is possible that the core is made of an iron–iron sulfide alloy, which would also produce a magnetic field but would have a density of only 5150 kg m^{-3}. It is also possible that the real moon is composed of only two layers – an iron core and a layer composed of a mixture of ice and rock – or that the proportion of rock and ice changes with depth or that... my point is that there are so many assumptions that could be wrong that it is important not to confuse the model with reality. We are not too likely to fall into this trap for Ganymede because of the simplicity of the model, and the model constructed by the Galileo science team (see Further Reading) was only a little more sophisticated than the one I have described here.

We have a little bit more information about Jupiter than Ganymede. Because Jupiter is essentially a large ball of gas, it is at least possible to look a little way into its interior. (The distinction between the interior and atmosphere is rather hazy for a planet that is mostly gas and I will also discuss the interiors of the gas giants in the next chapter on planetary atmospheres.) Observations from the Earth and from spacecraft have revealed something about the variation of temperature, composition and density in the upper layers of its atmosphere. Galileo also dropped a probe into Jupiter's atmosphere, and this probe transmitted information about the chemistry and other properties of the atmosphere until it was crushed by the pressure, which was unfortunately well before it reached some of the interesting depths described below. Well before the planet was visited by any spacecraft, it was already known from radio observations that Jupiter has a strong magnetic field, the strongest in the Solar System apart from the Sun. Finally, several of the J_n coefficients have been measured for Jupiter's gravitational field.

The two main difficulties in investigating the interiors of the giant planets are, first, that the increasing opacity with depth means we can only observe the upper layers of their atmospheres and, second, that the pressures at their centres are so great that our knowledge of how matter behaves under these conditions is often very poor. The pressure gradient within a planet is obtained from the principle of hydrostatic equilibrium (Chapter 1):

$$\frac{\mathrm{d}P}{\mathrm{d}r} = -\frac{GM(<r)\rho}{r^2} \tag{4.11}$$

When I used this equation in Chapter 1 to estimate the dependence of pressure on depth in the Earth, I made the assumption that density is independent of depth. This is not too bad an assumption for a solid planet, but it is clearly completely wrong for a gas giant. To use the principle of hydrostatic equilibrium to investigate the relation between pressure and depth in a gas giant, we need two additional things: an *equation of state* and a thermal model for the planet.

The most well-known equation of state is the relation between the pressure, density and temperature of a perfect gas:

$$\frac{P\langle\mu_A\rangle}{\rho} = RT \tag{4.12}$$

in which $\langle\mu_A\rangle$ is the mean gramme molecular weight in the atmosphere and R is the gas constant. This is only accurate if the molecules are far enough apart that the forces between them are negligible, which is not the case in Jupiter's dense atmosphere. In principle, scientists should be able to measure the equations of state for hydrogen and helium, the two main atmospheric gases, in the laboratory, but the pressure at the centre of Jupiter is much greater than the pressure that can be produced in any laboratory on Earth. Instead planetary scientists use equations of state that are extrapolated, using their knowledge of the atomic properties of there elements, from laboratory measurements at lower pressures.

The other unknown in Equation 4.12 is temperature, and to use the principle of hydrostatic equilibrium and the equation of state to determine how the pressure and density vary within a gas giant, it is also necessary to have some independent information about how the temperature varies within the planet. In an atmosphere in which convection is important, the temperature follows an *adiabatic temperature gradient*, a concept I will describe in the next chapter; it is likely that this is the case for Jupiter. With the equation of hydrostatic equilibrium, an equation of state and some knowledge of how the temperature varies within the planet, it is possible to determine how the pressure and density vary with depth. I will not attempt here to construct a detailed model of the planet, but suffice it to say that planetary scientists are able to use these basic ingredients to construct models of the pressure and density within a gas giant, all the way from the cloud layers, which is all we can actually see, down to the centre of the planet. As you might expect, the estimated pressure at the centre of Jupiter is extremely large (in one recent model ~20 times greater than the pressure at the Earth's centre). Some additional evidence that these models are broadly correct is that they provide a ready explanation of Jupiter's intense magnetic field.

The standard explanation of a planetary magnetic field is that it is the result of a planetary dynamo: convection currents in a conducting medium within the planet that amplify a seed magnetic field. The convection currents within a fluid iron core, for example, are the explanation of the Earth's field. At the high pressures deep inside Jupiter, the atoms are so close together that the electrons are no longer tightly held to individual hydrogen atoms, but can skip from one atom to another – the same behaviour seen in metals. The freedom of the electrons to move around means that this form of hydrogen conducts electricity, and convection currents in this *liquid metallic hydrogen* are a natural explanation of Jupiter's magnetic field. Figure 4.2 shows a schematic model of the planet. At the top is a layer of molecular hydrogen with the small amounts of other compounds that form the clouds, then there is the layer of liquid metallic hydrogen, and at the centre is a core of rock and ice.

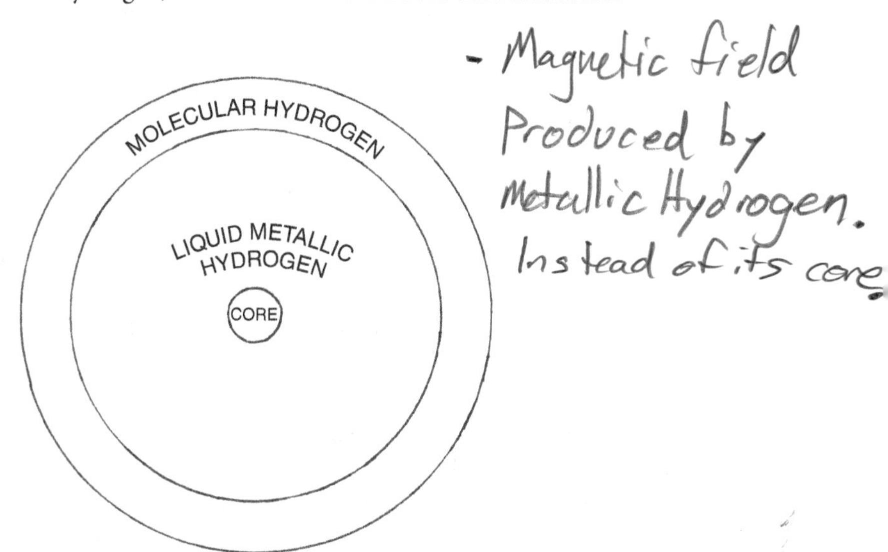

Figure 4.2 A model of the structure of Jupiter.

The most uncertain part of the model is the solid core. As with Ganymede, it is possible to use the value of the J_2 coefficient to estimate the moment of inertia of the planet, which for Jupiter is

$$\frac{I}{MR^2} = 0.254 \tag{4.13}$$

This is much less than the value for a planet with constant density (0.4) and also much less than the value for Ganymede, showing that the density of Jupiter increases even more strongly towards its centre. One recent model of Jupiter reproduces the low value of the moment of inertia by including a solid core of ice and rock with a mass that is ≈10 times the mass of the Earth, but which is only a small percentage of Jupiter's total mass (≈300 times the mass of the Earth). However, it is also possible to reproduce this low value for the moment of inertia with a model in which the gas density increases strongly towards the centre and in which there is no solid core at all. Our inability to decide whether or not there is a core is unfortunate because, as I will describe in Chapter 8, there are two competing theories for the formation of the giant planets: if one is true there should be a core; if the other is true there may well not be.

4.2 Mercury, Venus, Mars, Saturn, Uranus and Neptune

I described Ganymede and Jupiter in detail to show the limitations in our knowledge of planetary interiors. Before I turn to the planet about which we know most, I will summarize what we know about the interiors of the other planets.

Mercury has the second highest average density ($5440\,\mathrm{kg\,m^{-3}}$) of the planets in the Solar System, which suggests the planet has an iron core, which would also explain its magnetic field (Chapter 1). The moment of inertia of the planet is not known, because although there is a measurement of J_2, the tidal effect of the Sun (Chapter 6) has so slowed the planet's rotation and distorted its shape that there is no independent way of estimating the difference between the polar and equatorial moments of inertia. Nevertheless, if we assume that the planet consists of an iron core surrounded by a mantle of rock, then to explain the high average density, the iron core must comprise about 60 % of the mass of the planet.

Venus has a very similar mass and average density to the Earth, suggesting the structures of the two planets may be very similar. There is, however, remarkably little direct evidence about the planet's interior. Venus rotates very slowly and in the opposite direction to the other planets. The cause of this anomalous rotation is not known, but it throws doubt on the connection between the planet's rotational rate and its shape, which means there is not a reliable estimate of the planet's moment of inertia. One interesting difference between the two planets is that Venus does not have a strong magnetic field. Possible explanations of this difference are that either the iron core in Venus is solid or that for some reason there are no convection currents in the core.

The quality of the data for Mars is rather better than for Mercury and Venus. At least for this planet there is a good estimate of the moment of inertia. A simple model that is consistent with the two basic pieces of information about the planet – its moment of inertia and its average density – is that Mars contains a small iron core surrounded by a rock mantle.

Saturn's interior is probably rather similar to that of Jupiter, with any differences being the result of the difference in the planets' masses. Saturn's smaller mass means that the pressure at its centre is less than for Jupiter. The amount of hydrogen that is in the liquid metallic phase may therefore also be smaller, which may explain why Saturn has the weaker magnetic field.

The interiors of Uranus and Neptune are even more shrouded in mystery than those of the other planets. Although these planets are usually lumped together with Saturn and Jupiter, they actually have much lower masses (roughly one-twentieth the mass of Jupiter), and it is possible their interiors are quite different. Even their composition is uncertain. Observations of the upper layers of their atmospheres imply that elements other than the ubiquitous hydrogen and helium, such as carbon and sulfur, are more important in these planets than in the other two gas giants. Both planets have magnetic fields, and so there must be an electrically conducting fluid within each planet, but this cannot be liquid metallic hydrogen because the pressure in their interiors is too small for this form of hydrogen to exist. A possible explanation is that deep within the planets there are 'oceans' of either ionized water or some other ion. Both planets may contain a rocky core, but the data (as I'm sure by now you expect) is not good enough to say whether this is definitely so.

4.3 Why we know so much about the Earth

Let us now turn to our own planet. The reason we know so much about it is the seismic waves generated by earthquakes, which travel through the Earth and so contain information about its deep interior (there are also seismic waves that travel round the surface, but these are not very interesting for scientists unless you happen to be a scientist whose house is in an earthquake zone).

The two main types of seismic waves are P waves and S waves. P-waves (the P stands for 'primary') are pressure waves that travel through rock in the same way that sound waves are pressure waves that travel through air. They are longitudinal waves, which means that the vibration of the bits of rock occurs in the same direction that the waves travel. The velocity of the waves is given by

$$v_{\mathrm{p}}^2 = \frac{\left(K + \frac{4\mu}{3}\right)}{\rho} \qquad (4.14)$$

in which K is the bulk modulus of the rock, μ its shear modulus and ρ its density. The bulk modulus measures the resistance of a rock to being compressed and the shear modulus its resistance to being deformed by a twisting force. The shear modulus of a liquid is zero because a liquid has no shape to be deformed.

S waves (the S stands for 'secondary') are transverse waves, of which the best everyday example is a wave travelling along a string. In a transverse wave, the medium – the string or the rock – oscillates in a direction perpendicular to the one in which the wave is travelling. Transverse waves are only possible in a solid and are not possible in a liquid or gas. The velocity of an S wave is given by

$$v_S^2 = \frac{\mu}{\rho} \tag{4.15}$$

A comparison of the two equations shows that the P waves are faster than the S waves; the P waves from an earthquake therefore arrive at a seismic station before the S waves, which is why they are called *primary and secondary waves.*

It is usually between 15 and 30 minutes from when an earthquake occurs to when the P and S waves are detected at a seismic station on the opposite side of the Earth. The travel time contains valuable information because the wave velocities depend on the density and other properties of the rock through which the waves pass. I will show below how the travel times can be used to construct a much more detailed model of the Earth than the rudimentary models that are the only ones possible for the other planets, but I will first describe one of the most spectacular results from seismology, one that tells us something fundamental about the deep interior of our planet.

The types of waves detected at a seismic station depend in a remarkable way on the station's location relative to the earthquake's epicentre. Suppose I define the position of a seismic station by the angle between the lines joining the earthquake and the station to the centre of the Earth; the earthquake itself is therefore at zero degrees and the point on the opposite side of the Earth is at 180°. As we move round the Earth from the earthquake, seismic stations at angles less than about 110° detect both P waves and S waves; seismic stations between 110 and 145° *detect no seismic waves at all*; and seismic stations beyond 145° detect seismic waves again, but only P waves. We can explain these results with some basic high school physics.

Even if you no longer remember it, at some time during high school you undoubtedly learned Snell's law. This relates the abrupt change in direction (the refraction) that occurs when a ray of light travels from one medium to another to the different refractive indices in the two media, which in turn are the result of the different speeds of light in the two media. We can explain the results above by the refraction that occurs as seismic waves travel through the Earth. The detection of both P and S waves at angles less than 110° is easy enough to explain if the velocity of both kinds of wave gradually increases with depth; a seismic wave that sets off at a steep angle into the Earth will gradually curve upwards because of refraction and will eventually be detected by a seismic station on the Earth's surface (Figure 4.3).

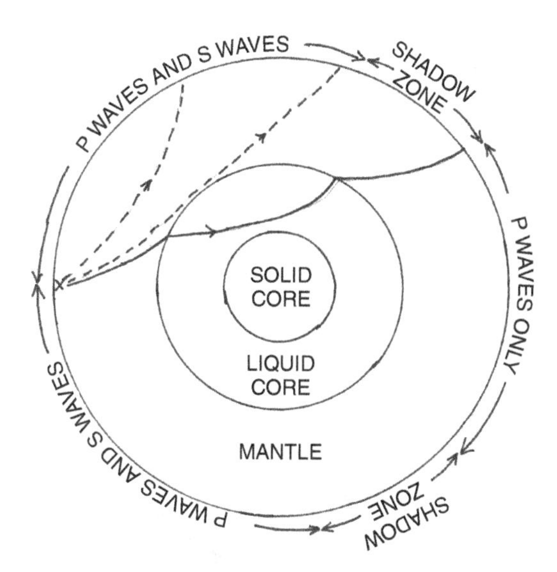

Figure 4.3 Seismic waves travelling through the Earth from an earthquake. The dashed lines show the paths of seismic waves that do not enter the core. The continuous line shows the path of a P wave that enters the core and re-emerges on the surface beyond the shadow zone.

The depth reached by a seismic wave depends on the angle of descent and, as the figure shows, if the angle of descent is sufficiently steep the seismic wave will reach the boundary between the core and the mantle. Now consider waves that travel at a slightly steeper angle of descent and do enter the core. Because S waves cannot travel through a liquid, the only possible explanation of the complete lack of S waves beyond 110° is if the Earth has a liquid core. The *shadow zone* between 110 and 145° can be explained if there is a sudden change in the P-wave velocity at the core–mantle boundary, in the sense that the velocity is lower in the core. A P wave that travels at a steep enough angle of descent to reach the core will be refracted at the core–mantle boundary; it will travel through the core and be refracted in the opposite direction when it travels back into the mantle – and the consequence of this double refraction is that it reaches the surface again at an angle of roughly 145° from the epicentre. The diagram in Figure 4.3 looks remarkably like the model of Ganymede that I discussed above, but the difference is that although this is still a model we have hard evidence that the Earth really is divided into these distinct layers (there is also evidence from the seismic data for a solid inner core).

The seismic data is a boon to the model-maker and it is possible to construct a much more sophisticated model of the Earth than is possible for any other planet. In the next few pages I will sketch how this is done, skipping over a few of the technical details (many whole books have been written about seismology).

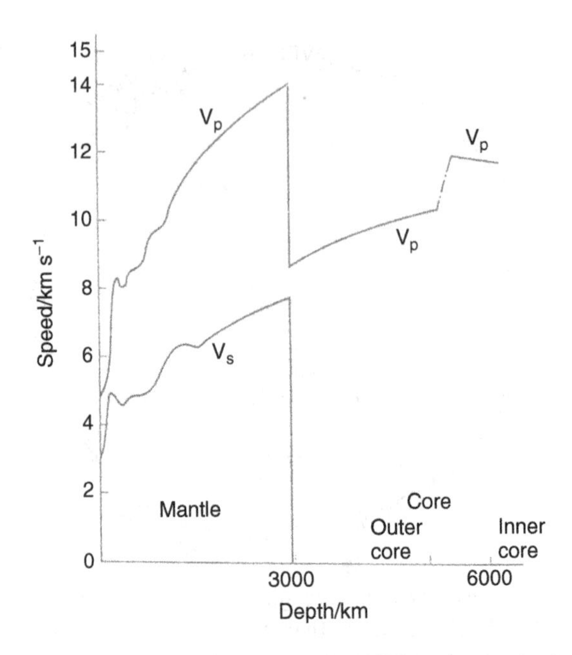

Figure 4.4 The variation with depth of the velocity of seismic waves.

The basic results recorded from an earthquake by the global array of seismic stations are the travel times for the P and S waves. The main detail I will skip is how the travel-time data are used to determine how the velocities of the P waves and the S waves depend on depth (Figure 4.4). This information by itself is enough to give some additional insights into the structure of the Earth; for instance, the increase in the P wave velocity at a depth of 5000 km is the evidence there is a solid inner core. By combining Equations 4.14 and 4.15 it is easy to show that

$$\frac{K}{\rho} = v_{\mathrm{p}}^2 - \frac{4}{3}v_{\mathrm{s}}^2 \tag{4.16}$$

and so once we know how v_{p} and v_{s} depend on depth, it is trivial to calculate how K/ρ depends on depth. Let us assume that we now have this information. I will describe one method, called the *Williams–Adams method*, for using this information to determine how density depends on depth within the Earth.

The starting point for any numerical calculation, which is what we are going to do, is to break the system we are going to model into discrete elements; in this case we will assume that the Earth is divided into a large number of spherical shells each of thickness Δr. In Chapter 1, I used the requirement that the Earth does not collapse under its own weight to show that the difference between the pressure at the top and bottom of a shell is given by the equation of hydrostatic equilibrium:

$$\frac{\Delta P}{\Delta r} \simeq -\frac{\rho GM(<r)}{r^2} \tag{4.17}$$

The bulk modulus is defined as the ratio of the change in the pressure on a substance to the fractional change in density that results from this change in pressure:

$$K = \frac{\Delta P \rho}{\Delta \rho} \tag{4.18}$$

By combining these two equations, we can eliminate ΔP to obtain the following equation:

$$\Delta \rho = -\frac{\rho^2 GM(< r)}{K r^2} \Delta r \tag{4.19}$$

We can now use this equation to calculate how the density varies within the Earth. Most numerical modelling is done on a computer, but the crucial part of the modelling is not writing the computer program but designing the *algorithm*. Once you have come up with the algorithm, you can either write the program or, if you have a lot of patience, implement the algorithm using a calculator and a piece of paper. These are the steps in the Williams–Adams algorithm:

Step 1: Using the information from the travel times, calculate the value of K/ρ at the centre of each shell.

Step 2: Start in the outermost shell. Assume that the density in this shell is the same as the density of rock on the surface of the Earth ($\approx$3000 kg m^{-3}). Assume that the mass in Equation 4.19 is the total mass of the Earth. We now know everything on the right-hand side of this equation. Use the equation to calculate the approximate change in density from the outermost shell to the shell beneath it.

Step 3: Calculate the density in the shell beneath the outermost shell. This is just the density in the outermost shell plus the change in density we calculated in step 2.

Step 4: Now calculate the mass interior to the second shell, which is approximately equal to the mass of the Earth minus the mass of the outermost shell (we are making the approximation that the mass of the second shell itself can be ignored). We now know again all the terms on the right-hand side of Equation 4.19 but this time for this next shell.

Step 5: Now calculate the approximate change in density between the second and third shells.

Step 6: Now calculate the approximate density in the third shell. . . .

But now the algorithmic snake has bitten its own tail. All we have to do to carry out the algorithm is to repeat the steps for each shell in turn, using Equation 4.19 to estimate the change in density from each shell to the one below, and then recalculating $M(< r)$ for the new shell by subtracting the masses of all the shells above. There are obviously lots of approximations in this recipe, but it is always possible to reduce the effect of these by increasing the number of shells into which the Earth is divided, which may not be practical with a pocket calculator but it is easy on a computer because even the cheapest personal computer can do billions of calculations each second.

The Williams–Adams method is a powerful one but not quite as powerful as it may at first seem, because it only works if there are no sudden changes in density within the Earth caused by a change in chemical composition or a change in phase. The sure sign that there is a chemical or phase change is a sudden change in the seismic wave velocity, such as the dramatic change in the P-wave velocity between the core and the mantle (Figure 4.4), and geophysicists have discovered many other smaller discontinuities. For example, in 1909 a Croatian geophysicist, Andrija Mohorovicic, discovered there is a sudden increase in seismic wave velocity at a depth of about 8 km under the oceans and 32 km under the continents. Geophysicists believe the cause of the *Mohorovicic discontinuity*, which thankfully they call the Moho for short, is a change in composition of the rock; they call the region above it the crust and the region below it the mantle[1]. Although the Williams–Adams method cannot cope with these discontinuities, using more sophisticated techniques and a certain amount of informed guesswork based on their knowledge of rocks, geophysicists have used seismic data to construct models of the Earth's interior of spectacular detail compared with the rudimentary models we have of the other planets. (The other information that geophysicists have about our planet is something I only have space to touch on parenthetically. Another consequence of a large earthquake, apart from the short-lived P and S waves, is that for days afterwards the whole Earth rings like a bell, and geophysicists also include the amplitudes and frequencies of these 'free oscillations' in their models.)

4.4 Why is the surface of the Earth such an interesting place?

Let us now finally turn to the question I posed at the beginning of this chapter. The answer to the subsidiary question of why the interiors of the planets are hot is to be found in the theories of how they were formed, which is something I will describe in detail in Chapter 8. For now, it is enough to state that according to the standard

[1]Rather confusingly, the terms crust and mantle are therefore not synonymous with lithosphere and asthenosphere. The change between the crust and the mantle is marked by the change in chemical composition revealed by the Mohorovicic discontinuity; the change from lithosphere to asthenosphere occurs when the pressure exceeds the strength of the rock. The one distinction comes from considering the chemistry of the rock, the other from looking at its physics.

theory of the origin of the solar system, the first large solid objects to form out of the protoplanetary nebula were *planetesimals*, which had sizes up to about 100 km in diameter; gravity then caused these planetesimals to eventually coalesce to form the planets. The planets should initially have been very hot because of the conversion of gravitational energy into heat that occurred in this process. We can estimate the amount of heat released in the formation of a planet in the following way.

The reader without any background in calculus should skip to Equation 4.24, which gives an estimate of the total amount of heat released during the formation of a planet. R_p *is the radius of the planet and* ρ *is its density.*

Let us assume that both the planetesimals and the *planetary embryo*, which is the name given to the largest planetesimal in the neighbourhood (Chapter 8), have the same density, ρ. The mass of the planetary embryo is given by

$$M = \frac{4\pi r^3 \rho}{3} \tag{4.20}$$

in which r is its radius. Let us suppose that some planetesimals collide with the planetary embryo. We will assume that the collisions are not vigorous enough to break up the planetary embryo, which therefore accretes the planetesimals onto its surface, growing in mass by

$$\delta M \simeq 4\pi r^2 \rho \delta r \tag{4.21}$$

which is equal to the mass of the accreted planetesimals. The amount of heat produced in these collisions is equal to the change in the gravitational potential energy of the planetesimals, which we will assume started out at infinity with zero velocity relative to the planetary embryo. The gravitational potential energy of 1 kg on the surface of the planetary embryo is $-GM/r$, and therefore the amount of energy released by the accretion of the planetesimals is given by

$$\delta E = \frac{GM\delta M}{r} = \frac{G16\pi^2 r^4 \rho^2 \delta r}{3} \tag{4.22}$$

In the right-hand side of this equation I have replaced M and δM by the expressions in Equations 4.20 and 4.21. This equation gives the amount of heat released when the size of the planetary embryo increases by δr, and the total amount of heat released during the formation of a planet is therefore given by the integral

$$E = \int_0^{R_p} \frac{G16\pi^2 r^4 \rho^2 \, dr}{3} \tag{4.23}$$

in which R_p is the current radius of the planet. The heat released in the formation of a planet is therefore

$$E = \frac{G16\pi^2 R_p^5 \rho^2}{15} \tag{4.24}$$

We can estimate the temperature of the planet immediately after it formed by putting this equal to the internal energy of the planet:

$$C_P T M = \frac{G 16\pi^2 R_P^5 \rho^2}{15} \tag{4.25}$$

in which T is the planet's temperature and C_p is its specific heat capacity. With a little rearrangement and using Equation 4.20 for the mass of the planet, we obtain the following equation for the temperature of the planet:

$$T = \frac{G 4\pi R_p^2 \rho}{C_p 5} \tag{4.26}$$

Using the specific heat capacity of basalt (800 J K^{-1} kg^{-1}), which is a plausible value to use for the average specific heat capacity of the Earth, I estimate that the temperature of the Earth immediately after its formation was about 41 000 K. This is well above the melting point of even the most refractory mineral ($\approx$1200 K), which implies that the Earth must then have been a ball of molten rock.

There is one thing I have left out of this calculation that might invalidate it. I have assumed that all the heat produced by the conversion of gravitational energy has been stored up in the planet, but it is possible that if the formation of the planet happened rather slowly, this energy might have been lost by radiation rather than being used to heat the planet's interior. More detailed models show that if the formation of the inner planets happened over a period longer than 100 million years, so much energy might have been lost by radiation that by the end of this period their interiors might still have been quite cold. At the moment we do not know very well how long it took the planets to form, but even if they were not molten initially, it seems likely that they would soon have been melted by another process.

The rocks in the Earth's crust contain many radioactive elements and the decay of these atoms releases energy and heats the Earth. The heating effect was greater 4.5 billion years ago because the abundances of radioactive elements were greater then. As the half-lives of radioactive elements are well-known (the half-lives of the important elements ^{235}Ur, ^{238}Ur, ^{232}Th and ^{40}K, for example, are 0.71, 4.5, 13.9 and 1.4 billion years), it would be easy enough to calculate the amount of heat released within the Earth at any time in the past, if only we knew the current abundances of all the radioactive elements. Unfortunately, although these are well-known for the crust, we obviously have no direct measurements of the abundances of radioactive elements in the core or mantle. Nevertheless, there is still one argument we can use to at least roughly estimate the importance of radioactive heating 4.5 billion years ago.

It is possible to estimate the rate at which heat is currently leaking out of the Earth using the Fourier heat law, which states that the heat flux Q (J s^{-1} m^{-2}) is proportional to the temperature gradient:

$$Q = -K_T \times \text{temperature gradient} \tag{4.27}$$

K_T is the *thermal conductivity*, which can be measured in the laboratory for rocks of different types. It is also possible to measure the temperature gradient in the Earth's crust by measuring the variation in temperature down a drill hole or as one goes down a deep mine. Therefore, by using Equation 4.27, it is fairly straightforward to estimate the rate at which heat is flowing out of the Earth's interior. By dividing the total heat flux from the Earth by its mass, one can estimate the current average rate of heat loss per kilogram of material, which is approximately $6 \times 10^{-12}\,\mathrm{J\,kg^{-1}\,s^{-1}}$. Although we cannot directly measure the radioactive heating rate in the Earth because we cannot measure the abundances of radioactive elements deep inside the planet, we can use meteorites as rough proxies for the Earth, since it is easy enough to measure the abundances of radioactive elements in them in the laboratory. The current rate of radioactive heating in one class of meteorite, the carbonaceous chondrites (Chapter 7), is $\approx 4 \times 10^{-12}\,\mathrm{J\,kg^{-1}\,s^{-1}}$, and so if the current radioactive heating rate in the Earth is broadly similar, it is not quite enough to replace the heat leaking out of its interior. Therefore today the Earth is cooling. However, from the half-lives of the different elements, it is easy to show that 4.5 billion years ago, the heating rate in meteorites must have been approximately ten times greater, and if this was also true of the Earth this would have been enough to melt the Earth's interior, if it was not molten already.

A hot interior is only part of the answer to the question of why the Earth's surface is such an interesting place. The rest of the answer lies in the different ways energy is transferred within the lithosphere and asthenosphere. In the asthenosphere, where on a geological timescale rock behaves like a fluid, heat is transferred mostly by convection. In the lithosphere, convection currents cannot occur and the only possible means of heat transfer are conduction and radiation. In silicate rocks, it is a mixture of the two: at low temperatures heat is mostly transferred by conduction (by vibrations that travel through the crystal lattice); at high temperatures, at which the rock becomes transparent to infrared photons, the heat is mostly transferred by radiation. To obtain the rest of the answer, we will have to estimate the efficiency of energy transfer within the lithosphere.

We may represent the thermal conductivity in Fourier's equation as the sum of a term arising from the lattice vibrations and a term arising from the transfer of photons:

$$K_T = K_L + K_R \tag{4.28}$$

In silicate rocks these depend on temperature in the following ways:

$$K_L = \frac{4.18 \times 10^2}{(30.6 + 0.21T)}\,\mathrm{J\,s^{-1}m^{-1}\,K^{-1}} \tag{4.29}$$

$$K_R = 0 \text{ for } T < 500\ \mathrm{K}$$

$$K_R = 2.3 \times 10^{-3}(T - 500)\ \mathrm{J\,s^{-1}\,m^{-1}\,K^{-1}} \quad \text{for } T > 500\ \mathrm{K} \tag{4.30}$$

Let us now estimate the maximum size of the region (L_{max}) in the lithosphere from which most of the energy can have leaked away by conduction and radiation in the 4.5 billion years since the formation of the Earth. The rigorous way to answer this question would be to use the thermal diffusion equation, a piece of maths that is beyond the scope of this book. However, we can obtain an approximate answer by using a powerful back-of-the-envelope technique called *dimensional analysis*, which is often used in physics when one is faced with a problem that is too complicated to solve by regular means.

Let us assume that there is some equation that connects the thing we want to estimate, in this case L_{max}, to the other parameters of the problem, which for the moment we will call X, Y and Z (there could, of course, be more than three of these):

$$L_{max} \sim f(X,\ Y,\ Z) \tag{4.31}$$

The first step in dimensional analysis is to use our scientific intuition, which is generally not too different from common sense, to write down a list of parameters that are likely to be important in the problem. In this problem, the first two parameters are fairly obvious: K_T and the age of the Solar System, τ. The energy contained in a cubic metre of rock is $C_p \rho T$, and so it seems possible that these three terms are also important. However, we can probably discard the temperature because Fourier's law shows that increasing the temperature of a lump of rock by a factor of 10 will increase the rate at which energy leaks out of the rock by the same factor, and so the percentage of the internal energy that leaks out each second will not change. We will therefore use only four parameters:

$$L_{max} \sim f(C_p, \tau, \rho, K_T) \tag{4.32}$$

The next step in dimensional analysis is to find the simplest equation linking these four parameters that has the same units as the left-hand side of the equation, which in this case are metres. There is a rigorous way of discovering this equation, but it is often possible to spot the answer, as it is in this case. By playing with the units of the different parameters (C_p – J kg^{-1} K^{-1}; τ – s; ρ – kg m^{-3}; K_T–J s^{-1} K^{-1} m^{-1}) on a scrap of paper, I can see that the units of the quantity $(K_T\tau)/(\rho C_p)$ are square metres. This is not quite right because the units on the left-hand side of Equation 4.32 are metres. This is an easy problem to fix by taking the square root:

$$L_{max} \sim \sqrt{\frac{K_T \tau}{\rho C_P}} \tag{4.33}$$

Using plausible values for the four variables on the right-hand side, I estimate a value for L_{max} of approximately 300 km. This shows that a small object, such as an asteroid, might have cooled down by conduction during the lifetime of the Solar System, but if conduction was the only process of heat transfer, the inner planets would have lost very little of their original heat content.

The inefficiency of heat transfer in the lithosphere is why the surfaces of planets are such interesting places. Convection carries energy efficiently up through the asthenosphere to the bottom of the lithosphere, but the energy cannot then readily escape by conduction through the lithosphere. The build-up of energy on its bottom surface creates stresses in the lithosphere. On the Earth, these stresses are released by the global system of plate tectonics, which has created many of the spectacular geological structures that we see on the surface of our planet. On Venus, these stresses may be released by volcanic activity produced from magma flowing through channels in the lithosphere, or they may build-up until the whole surface ruptures (Chapter 3). On the other two inner planets, of course, there is little current geological activity, but Mars and Mercury are much smaller than the other two planets and enough energy may now have leaked through their lithospheres that much of the stress on them has been removed. To the physicist, the geological cornucopia on a planetary surface – indeed the whole subject of geology – can be boiled down to the chance that in planets one physical process (convection) happens to be more efficient than another physical process (conduction).

Exercises

1 (calculus required) Show that if the density in a planet is a constant, the moment of inertia of the planet is 0.4 Ma^2, in which M is its mass and a its radius. (The moment of inertia of a thin spherical shell of radius r and thickness δr is $(8\pi\rho r^4\, dr)/3$).

2 (calculus required) Astronomers have discovered a new planet in the outer Solar System, which has a moment of inertia of 0.25 Ma^2, in which M is the planet's mass and a its radius. The planet is a long way from the Sun and so is likely to contain a large amount of ice. As a simple model of the planet, assume that it has a core with a radius of 0.5 a. If the outer part of the planet is made of ice, estimate the density of the core. (The moment of inertia of a thin spherical shell of radius r and thickness δr is $(8\pi\rho r^4\, dr)/3$; the approximate density of ice is 1000 kg m^{-3}.)

3 The average rate of heat loss through the Earth's surface is 0.074 J s^{-1} m^{-2}. Estimate the difference in temperature between the bottom of a mine 1 km below the surface and the surface. You may assume that the Earth's crust is made of silicate rocks.

4 On the assumption that the rate of heat loss given in Question 3 has been the same since the formation of the Earth, estimate the Earth's temperature shortly after it was formed. (Specific heat capacity of basalt rock: 840 J K^{-1} kg^{-1}; mass of Earth: 6×10^{24} kg; radius of Earth: 6378 km.)

Further Reading

Anderson, J.D., Lau, E.L., Sjogren, W.L. *et al.* (1996) Gravitational constraints on the internal structure of Ganymede. *Nature*, **384**, 541.

Kivelson, M.G., Khurana, K.K., Russell, C.T. *et al.* (1996) Discovery of Ganymede's magnetic field by the Galileo spacecraft. *Nature*, **384**, 537.

5

The atmospheres of the planets

Winter showers,
even the monkey searches
for a raincoat

Basho

5.1 The atmosphere of the Earth

The subject of this chapter is the only one that is impossible to ignore entirely in our everyday lives. I can ignore the existence of the planets, which are after all, unless one is interested in such things, just points of light in the sky, and unless I stub my toe on a rock or live in an earthquake zone it is easy enough to ignore the inner workings of the Earth and indeed the whole science of geology. But every day when I choose what to wear when I leave the house I have to consider the flow of gas in a planetary atmosphere and estimate the probability that the precipitation of liquid will occur. Ironically, this subject is also the one in which there sometimes seems to have been the least progress – at least judged by the accuracy of the weather forecast.

Given its importance to us, our planetary atmosphere is a surprisingly insubstantial thing. Let us start by investigating its structure, which is governed by our old friend, the principle of hydrostatic equilibrium (Chapter 1):

Those without calculus should skip to Equation 5.4, which describes how atmospheric pressure (P) depends on the height above the surface (z). The other terms in the equation are the mean molecular weight of the gases in the atmosphere ($\langle\mu_A\rangle$), the atomic mass unit (m_{amu}), Boltzmann's constant (k), temperature (T) and gravitational *acceleration* (g).

$$\frac{dP}{dz} = -\rho g \tag{5.1}$$

This looks a little different from Equation 1.10 in Chapter 1, but the two equations are essentially the same and have been derived in a similar way. In this equation, z is the height above the Earth's surface, g is the gravitational acceleration and P and ρ are the atmospheric pressure and density. The last two are also connected by the gas law:

$$\frac{P\langle\mu_A\rangle m_{amu}}{\rho} = kT \tag{5.2}$$

in which $\langle\mu_A\rangle$ is the mean molecular weight, m_{amu} is the atomic mass unit and k is Boltzmann's constant. We can combine the two equations to obtain the following one:

$$\frac{dP}{dz} = -\frac{P\langle\mu_A\rangle m_{amu}\, g}{kT} \tag{5.3}$$

Let us now make two assumptions: that g and T do not vary with height. (As you will see in a moment, the first of these is a very good approximation and the second one is OK because pressure changes with height much faster than temperature.) With these assumptions, the differential equation can be solved by separation of variables. The solution is

$$P = P_0 e^{-\left(\frac{\langle\mu_A\rangle m_{amu}\, gz}{kT}\right)} \tag{5.4}$$

in which P_0 is the atmospheric pressure at the surface.

This equation reveals that the pressure in the Earth's atmosphere drops surprisingly quickly. As an astronomer, one of my favourite places is Mauna Kea Observatory, which is on the summit of an extinct volcano on the Island of Hawaii (the 'Big Island'). The summit of Mauna Kea is 4205 m above sea level. With reasonable values for the temperature and the mean molecular weight, the equation shows that the atmospheric pressure at the observatory is only about 60 % of the pressure at sea level (the effects of this low pressure explain many of the dumb things that astronomers do while observing there). At the height of Mount Everest (8848 m) the pressure is only 33 % of that at sea level, which is why mountaineers in the Himalayas have to use breathing apparatus. Since the heights of these mountains are only a tiny percentage of the radius of the Earth ($\approx$6000 km), we can see at once that the Earth's atmosphere forms only a thin layer around it. It also shows that the assumption that the gravitational acceleration does not change significantly within the atmosphere is actually a very good one.

We can calculate the total mass of the atmosphere rather neatly. Since pressure is the force per unit area, the pressure at sea level is equal to the weight of the column of atmosphere above 1 m^2 of the surface, giving a very simple equation:

$$P_0 = m_{col}\, g \tag{5.5}$$

in which m_{col} is the mass of this column. The total mass of the atmosphere is therefore

$$M_{atmosphere} = \frac{4\pi R_E^2 P_0}{g} \tag{5.6}$$

in which R_E is the radius of the Earth. This is approximately 5.2×10^{18} kg, which is about one-millionth of the total mass of the Earth.

Before we turn to the other planets, let us consider in more detail what we know about our own atmosphere. Figure 5.1 shows how temperature depends on height. The complicated temperature profile is caused by the way in which the atmosphere is heated. The Earth's atmosphere is transparent over the wavelength range in which the Sun emits most of its radiation, which means that most of the Sun's radiation passes through the atmosphere and heats the surface. It is the infrared radiation from the warm surface (see Equation 1.6 in Chapter 1) that is the main energy source for the atmosphere. The atmosphere is warmest where it is closest to this heat source, and so in the lower part of the atmosphere, which is called the *troposphere*, temperature decreases with height. There is also, however, a second place where energy is injected into the atmosphere. Ozone consists of molecules in which there are three atoms of oxygen rather than the usual two, and the small amounts of ozone in the atmosphere absorb some of the Sun's ultraviolet radiation; there is a second temperature maximum (the *stratopause*) at a height of about 50 km, where most of this absorption occurs.

Atmospheric scientists divide the Earth's atmosphere into several layers. The troposphere is the most important one for daily life, because this contains most of the mass of the atmosphere and it is where the weather happens. The layer between the temperature minimum at the top of the troposphere (the *tropopause*) and the stratopause is called the *stratosphere*. There is a second temperature minimum (the *mesopause*) at a height of about 90 km, and the layer between this and the stratopause is called the *mesosphere*. The temperature increases with height above the mesopause because of the injection of solar energy by a variety of processes – this region is the *thermosphere*. The outer layer of the Earth's atmosphere, the *exosphere*, starts at the *exobase*. This is the height, between 500 and 1000 km, at which the density is so low that a fast-moving molecule is likely to escape into space before it collides with another molecule. The Earth's atmosphere consists of 78 % nitrogen (N_2), 21 % oxygen (O_2) and traces of other gases, the most important ones being water, carbon dioxide and argon (Table 5.1).

One of the most important parts of the atmosphere – and one of the most dramatic features of our planet as a whole (Figure 1.1) – are the clouds. The clouds are important in everyday life (am I going to get drenched when I cycle to work this morning?) and in the transport of water between the oceans and the land, but they are also important for the overall health of the planet as a habitat. The Earth's albedo depends on the fraction of its surface that is covered by clouds, because clouds

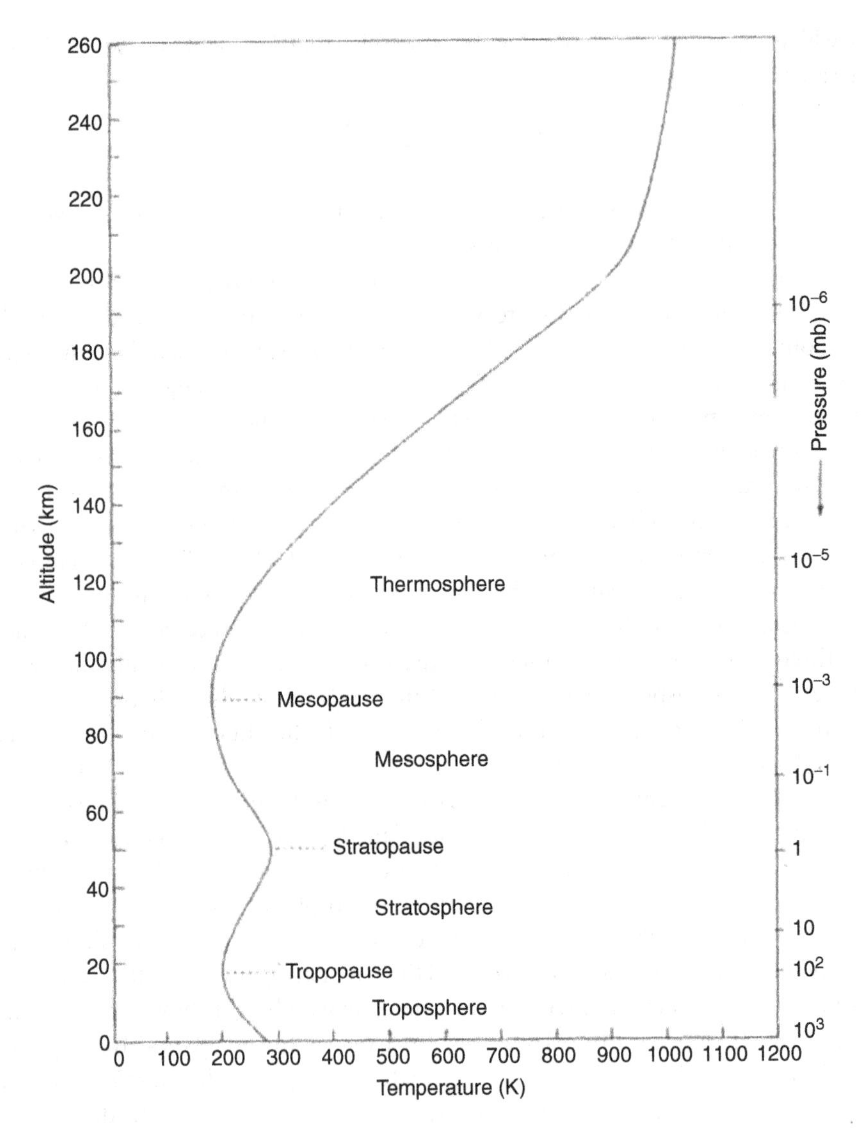

Figure 5.1 How the temperature in the Earth's atmosphere depends on height.

reflect more sunlight back into space than the oceans or the land. One of the biggest difficulties in estimating the temperature rise caused by the increasing amount of carbon dioxide in the atmosphere (Chapter 1) is our lack of understanding of the effect of the changing climate on the Earth's albedo.

Global warming is a serious business, but clouds are by their nature attractive things that are meant to lift the soul of a morning. Figure 5.2 shows some pictures of different types of cloud. These multifarious objects (the cloudspotter's

Table 5.1 The composition of the atmospheres of Earth, Venus, Mars and Titan.[a]

Constituent	Earth	Venus	Mars	Titan
N_2	0.78	0.035	0.027	0.90–0.97
O_2	0.21	0–20 ppm	–	–
Ar	0.009	70 ppm	0.016	48 ppm
H_2O	<0.03 (variable)	50 ppm	<100 ppm (variable)	–
CO_2	345 ppm	0.97	0.95	–
Ne	18 ppm	7 ppm	2.5 ppm	<0.01
O_3	10 ppm	–	–	–
He	5 ppm	12 ppm	–	–
CH_4	3 ppm	–	–	0.05
Kr	1 ppm	–	–	–
CO	–	50 ppm	700 ppm	10 ppm
NO	–	–	3 ppm	–
SO_2	–	60 ppm	–	–
H_2	–	–	10 ppm	0.002
C_2H_2	–	–	–	2 ppm
C_2H_6	–	–	–	10 ppm

[a]The value given for each chemical species is the fraction of the total number of particles (atoms or molecules) in the atmosphere belonging to that species, expressed either as a decimal fraction or in parts per million (ppm). I have only included constituents that are present at more than 1 ppm. The table is adapted from Table 4.3 of de Pater and Lissauer (2001), and I have added a few extra measurements for Titan from the Huygens lander (Niemann *et al.*, 2005).

guide – see Further reading – lists 28 different species) are seen at different levels in the atmosphere and are created in a number of different ways. Let us see how some of the prettiest clouds, the fluffy cumulus clouds that I can see out of my window this morning, are formed.

We will start by considering what happens to a small parcel of gas of unit mass as it moves up through the atmosphere if there is no heat exchange between it and the surrounding atmosphere (we will consider what might cause the gas to rise in a moment).

Those without calculus should skip to Equation 5.14, which gives the temperature gradient for an atmosphere in which there is no heat transfer between a slowly rising parcel of air and the surrounding atmosphere. The terms g and C_P are the gravitational acceleration and the specific heat capacity at constant pressure

The gas parcel must obey the first law of thermodynamics:

$$dQ = C_V\,dT + p\,dV \tag{5.7}$$

in which dQ is a small amount of heat that flows into the parcel from the surrounding atmosphere, dT and dV are the changes in the gas's temperature and volume, C_V

Figure 5.2 Four species of cloud. (a) Cumulus clouds, which are low-level (1000–2000 m) clouds and are formed by updrafts ('thermals') from the surface (see text); (b) low-level (0–1000 m) stratus clouds; (c) cirrus clouds, which are found at high levels in the atmosphere (3000–14 000 m) and contain ice particles rather than water droplets; (d) altocumulus clouds, which are found at heights of 1000–3000 m.

is the specific heat capacity at constant volume and P is the pressure of the gas. We are assuming that there is no heat flow into or out of the parcel and so $\mathrm{d}Q = 0$.

The gas in the parcel must also obey the gas law:

$$PV = \frac{RT}{M_\mathrm{m}} \tag{5.8}$$

in which M_m is the mass of one mole of gas.This equation must still be obeyed if the pressure, volume and temperature all change by small amounts, and so

$$(P + \delta P)(V + \delta V) = \frac{R(T + \delta T)}{M_\mathrm{m}} \tag{5.9}$$

By neglecting the very small term $\delta P \delta V$ and subtracting Equation 5.8 from Equation 5.9, we obtain the equation

$$P\,\mathrm{d}V + V\,\mathrm{d}P = \frac{R\,\mathrm{d}T}{M_\mathrm{m}} \tag{5.10}$$

in which I have written the small changes as infinitesimal quantities. By combining Equations 5.7 and 5.10 and assuming that no heat flows into or out of the parcel, we can obtain the equations:

$$0 = \left(C_V + \frac{R}{M_m}\right) dT - V\,dP = C_P\,dT - V\,dP \qquad (5.11)$$

I have obtained the second equation by using the relationship between the specific heat capacities at constant pressure and volume: $C_P - C_V = R/M_m$.

The pressure in the parcel of gas that is slowly rising through the atmosphere must always be the same as that of the surrounding gas – otherwise the parcel would rapidly expand or contract until the pressures are equalized. The atmosphere must be in hydrostatic equilibrium and so obeys the relation:

$$dP = -\rho g\,dz \qquad (5.12)$$

in which, as usual, ρ is the density, g is the gravitational acceleration and z is the altitude. Replacing dP in Equation 5.12 by Equation 5.11, we obtain

$$C_P\,dT + V\rho g dz = 0 \qquad (5.13)$$

and because $V\rho = 1$,

$$\frac{dT}{dz} = \text{temperature gradient} = -\frac{g}{C_P} \qquad (5.14)$$

Equation 5.14 is the dry adiabatic lapse rate, the rate at which temperature falls with height if the atmosphere is in hydrostatic equilibrium and if there is no exchange of heat between a parcel of gas rising though the atmosphere and its surroundings (*adiabatic* is the term describing a system like this in which there is no transfer of energy into or out of the system). With a correction for the latent heat released when water vapour condenses into clouds, Equation 5.14 describes the rate at which temperature falls with height in the troposphere ($\approx$10 K km^{-1}) rather well. Large parts of the atmosphere of all the planets have adiabatic temperature profiles like this. Why should this be?

The explanation is that convection currents tend to reduce any larger temperature gradient back to the adiabatic lapse rate. Let us consider the small parcel of gas again, but this time rising through an atmosphere in which the temperature is falling with height faster than the adiabatic lapse rate. Let us assume that the parcel rises quickly enough that no heat is exchanged between it and its surroundings, so that the temperature of the gas inside the parcel obeys Equation 5.14. As the parcel rises, its volume changes so that it is always in pressure equilibrium with its surroundings, but the temperature inside the parcel is greater than the temperature outside because of the different temperature gradients, which means (from Equation 5.2) that the density of the gas inside the parcel is less than the density outside. Archimedes' principle (Chapter 3) tells us the result: there is a buoyancy force, which causes the parcel to accelerate. Therefore an atmosphere like this is unstable. If the temperature gradient is greater than the adiabatic lapse rate, any upwards motion of gas will be strongly amplified by buoyancy forces, leading to a flow of energy from the bottom to the top of the atmosphere, thus reducing the temperature gradient. It is not surprising, therefore, that the temperature gradient in a planetary atmosphere is often close to the adiabatic lapse rate.

I have not so far answered the questions of why a parcel of gas should start moving in the first place or of why clouds should form. The two questions are related. On this February morning, as I write, many of the fields in the British countryside are still brown. Since a ploughed field absorbs sunlight particularly well, the air above these fields is warmer than the air above other places on the Earth's surface. Warm air rises because it has a low density (Archimedes' principle again), and so these fields produce updrafts or 'thermals'. The Earth's atmosphere contains water vapour, especially above a damp place like Britain. Let us now consider what happens to the water vapour in one of these updrafts. Figure 5.3 shows the phase diagram for water. At the Earth's surface, where I will assume the pressure and temperature in the atmosphere correspond to point A in the diagram, water is in the form of a gas (unless you have the bad luck to be in a stratus cloud – Figure 5.2). As air rises, its temperature falls, following the adiabatic lapse rate, and the pressure and temperature eventually reach point B in the diagram, which is on the line marking

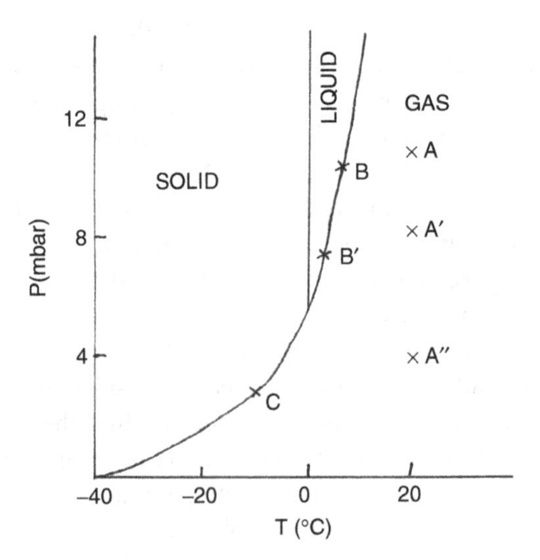

Figure 5.3 Phase diagram for water, showing the partial pressure of water vapour in the atmosphere plotted against temperature. The points A, A′ and A″ correspond to three different assumptions about the pressure and temperature of the water vapour in the Earth's atmosphere close to the surface. The points B, B′ and C show the points at which the water vapour in a rising parcel of air starting at these pressures and temperatures condenses or freezes.

the phase transition between gas and liquid – at which point the water vapour starts to condense into tiny droplets of water, and clouds form.

The cloudscape above our heads is the product of many factors. The bottoms of cumulus clouds are flat (Figure 5.2), because wherever the thermal starts on the Earth's surface, the water vapour condenses when the air reaches the height corresponding to the temperature necessary for the phase transition from gas to liquid. The sizes of the clouds, however, depend on the landscape below. The cumulus clouds produced from thermals off the large fields of the American Midwest look quite different from those above the small fields of the British countryside. The cloudscape reflects the landscape. The cloud level depends on the amount of water vapour in the atmosphere. Let us suppose that the partial pressure of water vapour at the Earth's surface is lower than before (point A' in Figure 5.3). As the parcel of gas rises, it will eventually reach the transition line between gas and liquid at point B', which is at a lower temperature than point B. In this situation, clouds will only start to form when the air reaches a higher level. Conversely, if the partial pressure of water at the surface is very high, the cloud level will be low, as it depressingly often is in Britain, the land of 'mists and mellow fruitfulness'. If the partial pressure of water vapour at the surface is extremely low (point A″), the rising parcel of gas will miss completely the transition line between gas and liquid but will eventually freeze into ice crystals (point C), forming high-level cirrus clouds (Figure 5.2).

5.2 The other planets

Let us now consider the other planets. Our knowledge of their atmospheres comes from a combination of spectroscopic observations and direct *in situ* observations made by spacecraft. The reason for the blue colours of Uranus and Neptune, for example, was revealed by optical spectroscopy: there are a large number of deep absorption lines at the red end of the optical wave band (Figure 5.4) caused by methane in the planets' atmospheres, which means the reflected sunlight that we see is very blue. Table 5.1 shows the composition of the atmospheres of the three inner planets with significant atmospheres (Mercury does not have an atmosphere worth mentioning) plus Titan, the only moon with a substantial atmosphere. I have listed the elements and compounds in order of their importance on Earth. The table shows that the Earth's atmosphere is very different from the atmospheres of Venus and Mars, being dominated by oxygen and nitrogen. Nitrogen is only a minor constituent of the atmospheres of Venus and Mars and oxygen is hardly present at all. Their atmospheres are dominated by carbon dioxide, which is only fifth in importance on the Earth. Titan's atmosphere like the Earth's is mostly nitrogen, but it also contains a significant amount of methane and other organic compounds, which makes it very different from the atmospheres of the planets.

We know less about the composition of the atmospheres of the giant planets because the only spacecraft that has made *in situ* measurements was the Galileo

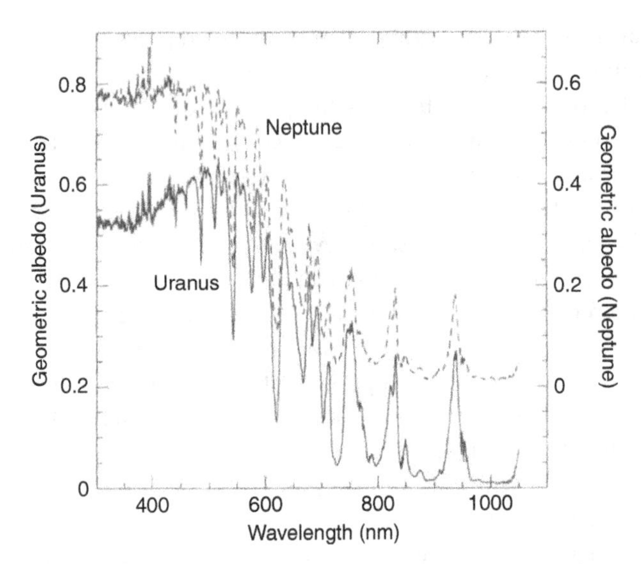

Figure 5.4 Optical spectra of Uranus and Neptune. Neither planet reflects much sunlight at wavelengths greater than 600 nm because the methane in their atmospheres has many absorption features beyond this wavelength (adapted from Figure 4.13 of *Planetary Sciences* by de Pater and Lissauer, 2001).

probe, which descended into Jupiter's atmosphere until it was crushed by the pressure. For these planets, therefore, we are reliant on remote observations, which are themselves limited by the dense atmospheres of these planets and the cloud layers. The clouds are mostly opaque to optical and ultraviolet radiation, and it is only possible to investigate the deeper levels of the atmospheres from observations in the far-infrared, submillimetre and radio wave bands. The atmospheres of the giant planets seem to be almost entirely composed of hydrogen and helium, with only tiny amounts of other gases, the main ones being methane, water, ammonia and hydrogen sulfide – gases which condense or freeze at the temperatures and pressures in the atmospheres of the giant planets and which form their clouds. All these other gases comprise only about 1 % (by number of particles) of the atmospheres of Jupiter and Saturn. The only gas, apart from hydrogen and helium, which is at all significant in the atmospheres of the giant planets is methane, which constitutes about 3 % of the atmospheres of Uranus and Neptune, and which gives these planets their colour.

Our knowledge of the temperatures within the atmospheres of the giant planets also comes almost entirely from spectroscopy, in particular from the brightness of the spectral lines. As a simple example of how this is done, suppose that the continuum emission is from deep inside the planet and the spectral lines are from an upper layer in its atmosphere: the spectral lines will be absorption lines if this layer is cooler than the layers beneath and emission lines if this layer is warmer. Both spectroscopy and the measurements by the Galileo probe suggest that the temperature in the gas giants varies with depth following the *adiabatic temperature profile* seen in the troposphere of the Earth.

The atmospheres of all the other planets (and Titan) contain clouds, although none, with the possible exception of Jupiter, as attractive as those on the Earth. The clouds on the other planets are also the consequence of compounds that can exist in more than one phase within the temperature range on the planet. Mars is most similar to the Earth because its low-level clouds ($\sim$10 km) are also composed of water, but like the cirrus clouds on the Earth, these clouds are composed of ice crystals rather than water droplets because of the low partial pressure of water vapour in the Martian atmosphere. On Mars, however, there are also clouds composed of carbon dioxide, which form at a height of $\approx$50 km, where the temperature is low enough for the gas to freeze. On Venus, the thick clouds that shroud the surface are composed of droplets of sulfuric acid formed by the photochemical action of the Sun's ultraviolet light on the carbon dioxide, sulfur dioxide and water vapour in the atmosphere. About the clouds on the outer planets much less is known, because for these we have to rely on remote sensing (the clouds themselves are part of the problem, the upper cloud layers obscuring the lower ones). However, it seems likely that there are several layers of clouds on each planet, composed of ammonia, hydrogen sulfide, methane or water – all of which should freeze or condense at different levels in the atmospheres.

The only moon with a substantial atmosphere is Titan. Water on Earth has the special property that it exists in all three phases – gas, liquid and solid – and scientists have long realized that methane might have the same property on Titan, raising the possibility of methane lakes and methane ice, with methane snow and rain falling from the sky. The recent results from the Cassini mission have largely confirmed this suspicion. The Huygens probe, which descended onto the surface of Titan, found evidence for the evaporation of methane from the surface, suggesting there is a methane cycle like the water cycle on the Earth. The most spectacular discovery, however, was made with the radar experiment on the Cassini mother-ship overhead. The scientists running this experiment discovered 75 areas on Titan's surface that reflect radio waves very poorly, as do the lakes and oceans on the Earth. The radar-dark areas on Titan also look remarkably like large lakes (Figure 5.5). As usual, the NASA publicity machine has slightly over-cooked the evidence, because in this false-colour image of the radar data somebody has thoughtfully painted the radar-dark areas blue in case one misses the point that they are supposed to be lakes. Nevertheless, it is difficult to see what else they could be. The patches range in size from 3 to 70 km and, if they are lakes, they are the first ones discovered outside the Earth.

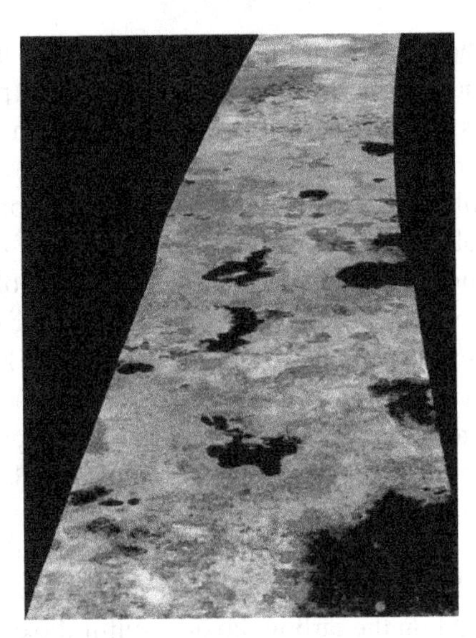

Figure 5.5 A false-colour image of the surface of Titan made using radar data from Cassini. Areas that reflect radio waves poorly are shown as blue. A colour reproduction of this figure can be seen in the colour section, located towards the centre of the book (Courtesy NASA)

5.3 The weather on the Earth and elsewhere

The atmospheres of the planets are all very different, but as with some of the other properties of the planets, one can sometimes see underlying principles that lie behind this diversity. One nice piece of physics connects the bands seen in the image of Jupiter (Figure 5.6) and the direction of the clouds I can see through my window scudding across the sky.

The clouds are coming from the west, which is the prevailing wind direction in this part of Wales and indeed in the UK as a whole. The prevailing winds are from the west between the latitudes of 30° N and 60° N, from the east between latitudes of 30° N and 30° S, and from the west between 30° S and 60° S. This wind pattern was one of the two reasons (the other was money, of course) for the ghastly triangular trade that developed in the eighteenth century: ships sailed from Britain to the west coast of Africa, where they exchanged cheap manufactured goods for slaves; then sailed to North America, aided by the 'trade winds', where they sold the slaves; and then, in the final leg of the triangle, carried tobacco and rum from the slave plantations back to Europe, propelled by the 'westerlies'. Why is there this distinctive pattern of winds?

On the terrestrial planets, there is only one possible energy source: the Sun. Edmund Halley at the end of the seventeenth century was the first to realize that the ultimate cause of the winds is that the Sun heats the equator more strongly than

Figure 5.6 Image of Jupiter taken by Voyager 2. The Great Red Spot is just below the centre of the image (courtesy NASA).

the poles. The air at the equator is therefore hotter and has a higher pressure than the air above the poles, which creates a simple circulatory pattern: hot air rises at the equator and flows at high altitudes towards the poles; cold air flows back at low altitudes from the poles to the equator. Of course, this is not what we see, which is a pattern of east – west winds rather than north–south ones. The reason why we see an east–west pattern of winds was first realized by an English lawyer called George Hadley in the next century.

The reason there is not a wind blowing directly from the North Pole (fortunately) is that the Earth is rotating. Anyone who has ever watched the TV weather report knows that winds do not actually blow in a straight line. This is easiest to understand by thinking about a children's roundabout. Suppose a child on the roundabout has just thrown a ball. Someone standing on the ground beside the roundabout sees the ball move in a straight line, but someone on the roundabout itself sees it move in a curve. Linear motion in an inertial reference frame is transformed into a curve in a rotating reference frame, and planets like roundabouts are rotating reference frames.

We can investigate in more detail how the winds blow on the Earth by using the law of conservation of angular momentum. Let us suppose that a wind at latitude θ has a velocity component towards the north of v_N and towards the east of v_E. The angular momentum of a parcel of air of unit mass around the Earth's axis is given by

$$L = \Omega R^2 \cos^2 \theta + v_E R \cos \theta \tag{5.15}$$

in which R is the Earth's radius and Ω is the rate at which it rotates. The first term in the equation is the angular momentum that any parcel of air possesses, even if there is no wind at all, because of the rotation of the Earth; the second term is the extra angular momentum if there is a wind blowing. Now let us suppose there is a wind at the equator blowing towards the north.The second term is therefore zero. As the air travels to higher latitudes, the first term in the equation gradually decreases. But because of the law of conservation of angular momentum, the total angular momentum is a constant and so the second term in the equation must increase – the wind therefore veers towards the east. The flow of cold air at low altitude from the poles towards the equator, which is expected from Halley's hypothesis, therefore veers in the opposite direction, which explains why the prevailing winds at the equator are towards the west. In honour of the person who first realized that a north–south circulation of air might actually produce an east–west pattern of winds, the circulation of air from the equator to the poles is now called *Hadley circulation*. In our rotating reference frame, the acceleration of the air is explained by the fictitious *Coriolis force*.

The reader who is paying attention will realize that I haven't actually explained why the direction of the prevailing winds reverses with latitude. Unfortunately, there is no simple explanation. The flow of air in a planetary atmosphere is an extremely complex process because of many other effects such as turbulence, which is why meteorologists invest lots of money in large computers to simulate the flows of air (and why the weather forecast is so often wrong). Although it is not possible to predict the end-result of all these effects without a computer simulation, what seems to happen is that the Coriolis force disrupts the circulation of air between the pole and equators so there are often several north–south circulatory flows of air. On the Earth in the northern hemisphere there are three of these *Hadley cells* (Figure 5.7). The effect of the Coriolis force on the low-altitude flow in each of

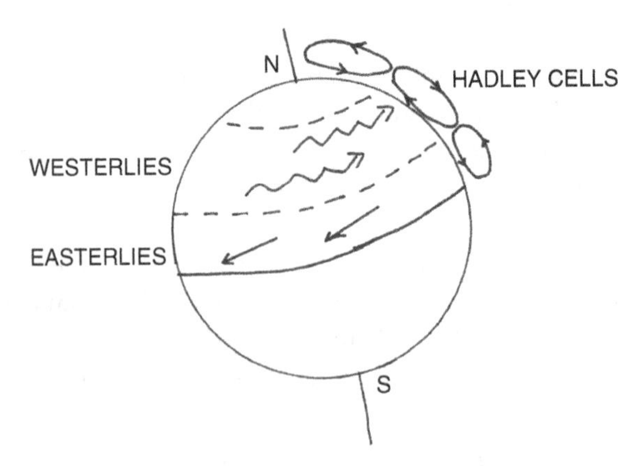

Figure 5.7 The three Hadley cells in the Earth's northern hemisphere. The prevailing wind directions are the result of the Coriolis force acting on the low-altitude flow in each Hadley cell.

these cells explains nicely the direction of the prevailing winds at different latitudes. Venus rotates very slowly and so there is only a weak Coriolis force, which explains why its atmosphere appears to contain only a single Hadley cell.

This idea connects the weather on Jupiter to the weather on Earth. Spectroscopy shows there is an east–west pattern of winds on Jupiter, with the wind direction reversing several times as one moves from the poles to the equator, and that this pattern is correlated with the bands seen in pictures of Jupiter (Figure 5.6). This suggests that circulatory cells within its atmosphere are the cause of Jupiter's distinctive appearance (there are more of these cells on Jupiter because it is rotating faster). The winds on Jupiter are very strong and, according to computer simulations, such strong counter-flowing streams of air should generate vortices at their boundaries, which change constantly as new ones are formed and old ones disappear or combine with others – some of these can be seen clearly at the boundaries between bands in Figure 5.6. One difference between the weather on Jupiter and terrestrial weather was discovered by the Galileo probe as it descended into its atmosphere; on Jupiter the strength of the winds increases with depth. This suggests that on Jupiter there may also be an internal energy source for the weather, not just the differential heating effect of the Sun.

There is another connection between weather on the two planets. As anyone who has listened to a weather forecast with even half an ear knows, weather on Earth is dominated by high-pressure and low-pressure systems. The contours on weather maps are isobars, lines joining places in the atmosphere with the same pressure. One might expect that air would flow towards a low-pressure system and away from a high-pressure system, but the effect of Coriolis force is that winds blow along the isobars (the proof is beyond the scope of this book). Winds that blow around low-pressure regions are called cyclones; winds that blow around high-pressure regions are anticyclones. The effect of Coriolis force is different in the two hemispheres, with anticyclones blowing clockwise in the north and anticlockwise in the south and the reverse for cyclones. The most spectacular example of weather on another planet is the Great Red Spot on Jupiter (Figure 5.6). Winds blow around its centre in an anticlockwise direction, travelling a complete circuit in about 6 days, making the Great Red Spot, since it is in Jupiter's southern hemisphere, an anticyclone. A big difference, however, is in the duration of the weather systems on the two planets. The Great Red Spot has been in existence since at least 1665 when it was first noticed by Giovanni Cassini. When bad weather sets in on Jupiter it sets in for a long time.

5.4 The origin and evolution of planetary atmospheres

Let us now consider where the atmospheres come from. The origin of the atmospheres of the outer planets is so bound up with the origin of the planets themselves, which is described in Chapter 8, that here I will only discuss the inner planets. There are two possibilities. Either the atmospheres are 'primary atmospheres', consisting of gas accumulated from the solar nebula while the planets themselves were forming or they are 'secondary' atmospheres, which were produced after the formation epoch of the planets. There is a simple argument against the first idea. The solar nebula out of which the planets were formed (Chapter 8) must have contained the same abundances of elements as in the Solar System today (Table 1.2). If the atmospheres are primary atmospheres, they should therefore contain similar proportions of elements, which a comparison between Tables 1.2 and 5.1 shows they clearly don't. It is easy to explain away some of the differences. The lack of hydrogen and helium in the Earth's atmosphere today can be explained by the lightness of their atoms; any original hydrogen or helium would have gradually escaped from the Earth's gravitational field over the last 4.5 billion years (see below). The smoking gun is the fifth most abundant element in the Solar System: neon. Despite its abundance in the Solar System as a whole, there is very little neon in the atmospheres of any of the inner planets (Table 5.1), but neon atoms are quite heavy and so do not easily escape from the planets' gravitational fields, and they are also chemically inert and so cannot have been removed in any other way. Therefore the atmospheres of the inner planets are almost certainly secondary atmospheres.

The atmospheres of the inner planets were probably formed by the gradual release of gases from their interiors. It seems likely that the inner planets were formed from the coalescence of smaller 'planetesimals' (Chapter 8), some of which can still be seen as asteroids and comets in the Solar System today. Comets, in particular, often contain large reservoirs of ices, including frozen water, carbon dioxide and methane (Chapter 7), and ice-rich planetesimals incorporated in the planets were probably the source of their atmospheres. These might have been produced slowly as volcanic activity released gases from the planets' interiors – as still happens on the Earth today – but it is also possible they were formed very quickly when some final ice-rich planetesimal collided with a planet and was completely vaporized by the heat of the collision.

We do not know the composition of the Earth's original atmosphere. The two main possibilities are that either it was a *reducing atmosphere* rich in compounds containing hydrogen, such as methane (CH_4), ammonia (NH_3), water (H_2O) and hydrogen sulfide (H_2S) or it contained mostly compounds with no hydrogen, such as carbon dioxide (CO_2), nitrogen (N_2) and sulfur dioxide (SO_2) – an *oxidizing atmosphere*. Although the latter is more similar to the atmospheres of the planets today, we cannot be sure which it was because of several processes that have transformed the atmospheres of the planets during the last 4.5 billion years.

The most important process is the same one that accounts for why only big objects have atmospheres: the escape of gas particles into space. If a gas particle is in the planet's exosphere, where the chance of it colliding with another particle is very low, it will escape into space if its velocity is greater than the planet's escape velocity. The velocity distribution for a gas in thermal equilibrium is given by the Maxwell distribution, and thus the fraction of gas particles with velocities higher than the escape velocity, v_{esc}, is given by

$$p(>v_{esc}) = \int_{v_{esc}}^{\infty} N \left(\frac{2}{\pi}\right)^{\frac{1}{2}} \left(\frac{m}{kT}\right)^{\frac{3}{2}} v^2 e^{-\frac{mv^2}{2kT}} \, dv \qquad (5.16)$$

in which N is the number density of particles and m is the mass of each particle.

In reality, things are a bit more complicated, because the gas in the exosphere is not necessarily in thermal equilibrium and there are other processes (at least six) that affect the escape of gas into space, including the dissociation of molecules by ultraviolet photons and the interaction of ions with the solar wind, but there are two safe conclusions we can draw from this equation. The mass dependence shows that the lighter a particle, the greater the chance it will escape from the planet, and the fact that the lightest of all elements, hydrogen, is lost so readily from a planet probably explains why, if the planets once had reducing atmospheres, they now have oxidizing atmospheres. The dependence on v_{esc} shows that the rate of gas loss is greater for small objects and explains why the small objects in the Solar System only have tenuous atmospheres.

This process may also explain some of the differences between the atmospheres of the inner planets. Venus and the Earth, similar in mass and density, are very different in many other ways. One difference I have not mentioned so far is that while the Earth is the 'blue planet' there appears to be very little water anywhere on Venus. It is possible that Venus never had any water – possibly it formed from planetesimals that contained very little ice – but it is also possible that it lost its water because of the photodissociation (by solar photons) of water molecules in the atmosphere into oxygen and hydrogen atoms, with the latter gradually leaking away into space. The escape of atmospheric gases into space might also explain Mars' vanishing atmosphere. I argued in Chapter 3 that the evidence for running water over the Martian surface showed that Mars must once have had a dense atmosphere. Since Mars has a much lower mass than the two other planets, it is possible that this atmosphere was gradually lost into space.

An alternative explanation is another process that has gradually transformed the atmospheres of the inner planets: the Urey weathering reaction, in which carbon dioxide dissolved in water reacts with silicates in rocks to form calcium carbonate (Chapter 3). On the Earth, as the result of this process, most of the carbon dioxide that was originally in the atmosphere is now locked up in carbonate rocks, which explains why Venus has a dense carbon dioxide atmosphere but the Earth does not. On the Earth, however, there is also a reverse process, because carbon dioxide is returned to the atmosphere through volcanic activity when tectonic plates are forced down into the asthenosphere, where the heat breaks down the carbonate rocks. The result of this *carbonate–silicate cycle* is to keep the amount of carbon dioxide in the atmosphere fairly constant (it *is* growing but only because of human activity) and it may also explain why the temperature of the Earth has stayed roughly constant for the last 4.5 billion years, despite the Sun increasing in brightness by 30 % during this time (Chapter 9). On Mars, one side of the cycle is no longer happening because there is little current volcanic activity. Therefore it is possible that the missing atmosphere is locked up in carbonate rocks, although the failure of Mars Express to find any spectroscopic evidence of these is an argument against this idea (Chapter 3).

The most interesting transformative process is life. Not only the oxygen in the Earth's atmosphere but also the trace gases methane and ozone are the result of this process. In contrast, there are no clues from the atmospheric composition of Mars and Venus that there has ever been life on these planets. This transformative process has the interesting consequence that life has had to evolve to cope with the changes in the atmosphere it has caused, for example the evolutionary adaptations necessary to cope with the flooding of the atmosphere by oxygen from photosynthesis 2 billion years ago. In Chapter 9 we will consider why the Earth is alone among the inner planets in being a hospitable place for life and why, despite large variations in the composition of its atmosphere, it has remained a hospitable place for life for at least 3 billion years.

Exercises

1 Using the information in Tables 1.1 and 5.1, estimate the height above the surface of Titan at which the pressure has fallen to 20 % of its surface value. Suggest one reason why Titan has a dense atmosphere, but Mars, which has a much larger mass than Titan, does not (mass of Titan: 1.3×10^{23} kg; radius of Titan: 2.6×10^{6} m; molecular weight of nitrogen: 28).

2 If the temperature at the Earth's surface is 20°C, by referring to Figure 5.3 estimate the height at which the clouds start. You should assume the dry adiabatic lapse rate and assume that the partial pressure of water vapour is 7 mbars and is roughly constant with height (specific heat capacity of air: $\approx$1000 J kg^{-1} K^{-1}; gravitational acceleration: 9.8 m s^{-2})

3 The atmospheric pressure on Venus is about 700 times that on the Earth. Calculate the mass of Venus' atmosphere. If this atmosphere was locked up in carbonate rocks, estimate the average thickness of the carbonate rocks that would cover the Venusian surface (radius of Venus: 6052 km; surface gravity on Venus: 8.9 m s^{-2}; atmospheric pressure on Earth: 10^{5} N m^{-2}; density of calcium carbonate rocks: $\approx$2000 kg m^{-3}; atomic weights of calcium, carbon and oxygen are 40, 12 and 16).

4 (challenging) The magnitude of the Coriolis force acting on a unit mass of air is $2\Omega v \sin(\delta)$, in which Ω is the angular velocity of the Earth, v is the wind velocity and δ is the latitude. Suppose that the wind is part of an anticyclonic storm system and is travelling at 100 km hour^{-1} in a circle around a high-pressure region in the Earth's atmosphere. Estimate the radius of the storm system if it is at a latitude of 50°.

Further Reading and Web Sites

Adamkovics, M., Wong, M.H., Laver, C. and de Pater, I. (2007) Widespread morning drizzle on Titan. *Science*, **318**, 962.

de Pater, I., and Lissauer, J.L. (2001) *Planetary Sciences*, Cambridge University Press, Cambridge.

Cloud Appreciation Society web site http://www.cloudappreciationsociety.org (accessed 18 September 2008)

Cassini–Huygens web site http://saturn.jpl.nasa.gov/home/index.cfm (accessed 18 September 2008)

Niemann, H.B., Atreya, S.K., Bauer, S.J. (2005) The abundances of constituents of Titan's atmosphere from the GCMS instrument on the Huygens probe *Nature*, **438**, 779–84.

Stofan, E.R., Elachi C., Lunine J.I. *et al.* (2007) The lakes of Titan. *Nature*, **445**, 61.

6 The dynamics of planetary systems

If I have ever made any valuable discoveries,
it has been owing more to patient attention
than any other talent

Isaac Newton

6.1 Laws of planetary motion

The meandering paths of the planets across the sky (the word planet comes from the Greek for wandering star) have fascinated people since at least the time of the Babylonians. It is the explanation of this motion that was the greatest triumph of the scientific revolution of the sixteenth and seventeenth centuries. With the twenty–twenty hindsight of history it is easy to think that this was actually rather a simple problem, but it is important to remember that the motion of the planets is quite complicated. Most of the time an outer planet (beyond the orbit of Earth) travels across the sky from west to east, but every now and then it will abruptly go into reverse, and the challenge of thinking of a simple theory to explain 'retrograde motion' defeated astronomers for 2000 years.

The problem was solved by one observer and two theorists. The observer was the Danish aristocrat, Tycho Brahe. In the late sixteenth century, Tycho built the first large astronomical observatory on an island off the Danish coast: Uraniborg, the 'castle of the sky'. For over 20 years Tycho and his team of assistants measured the positions of the planets with an accuracy almost 10 times better than previous

Planets and Planetary Systems Stephen Eales

measurements, which is startling considering this was still the generation before the invention of the telescope.

One of Tycho's assistants, for a time, was Johannes Kepler. Half astronomer and half astrologer, Kepler's perennial goal, whether he was being an astronomer or an astrologer, was to find a simple pattern behind the complex motions of the planets. On Tycho's death, Kepler inherited (or walked off with them, depending on who you talked to) Tycho's accumulated measurements of the position of Mars – a priceless resource for a theorist interested in planetary motion. He realized he could explain Tycho's results if a planet obeys three laws:

1) It follows an elliptical path with the Sun at one focus.

2) The line joining the planet and the Sun sweeps out area at a constant rate.

3) The square of the time the planet takes to go round the Sun, P, is proportional to the cube of the semi-major axis of its orbit, a: $P^2 \propto a^3$.

The complex motion we see in the sky is produced from these simple laws, now known as *Kepler's laws*, because we are observing the planets' motions, not from a position perched above the Solar System, but from the surface of one of them. For example, Kepler's third law shows that the Earth travels around the Sun in less time than Mars, and Mars' retrograde motion occurs when the Earth overtakes it (Figure 6.1).

These are descriptive laws. They show what the planets do, but not why they do it. Isaac Newton showed why the planets obey Kepler's laws. He realized that these are a consequence of a force that exists between the planets and the Sun, and indeed between any pair of objects, which is given by the equation:

$$F = \frac{GM_1 M_2}{d^2} \tag{6.1}$$

in which M_1 and M_2 are the masses of the two objects and d is the distance between them. It is not actually that easy to show that Kepler's first two laws are a result of this equation (Appendix 2). In contrast, it is quite easy to show that Kepler's third law follows from Newton's law of gravity, at least for circular orbits (the orbits of the planets are very close to being circles). Suppose that the planet is moving in an orbit of radius a with an angular velocity Ω. The centripetal force necessary to keep the planet in its orbit is $M_P\Omega^2 a$, in which M_P is the mass of the planet. The centripetal force is provided by the gravitational attraction between the planet and the Sun, and so

$$M_P\Omega^2 a = \frac{GM_P M_s}{a^2} \tag{6.2}$$

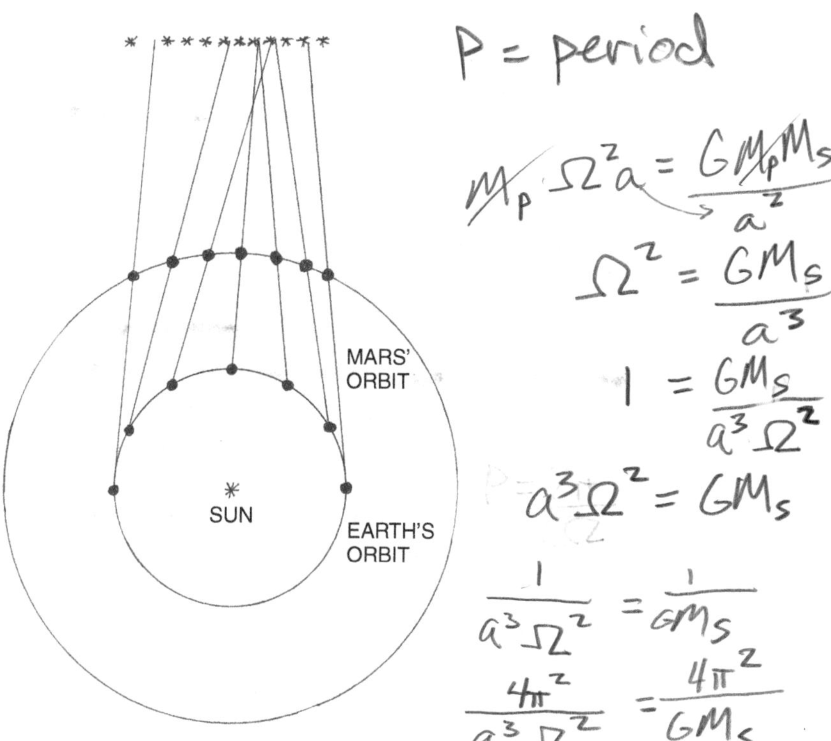

Figure 6.1 How Kepler's third law explains retrograde motion. Each straight line links the positions of Earth and Mars at the same time – the line's projection onto the celestial sphere shows where Mars appears against the stars as seen by an observer on the Earth. Because Mars takes longer to go round the Sun, for a brief period it appears to travel backwards in the sky.

The period of the planet's orbit, P, is equal to $2\pi/\Omega$. With a little rearranging, Equation 6.2 becomes

$$P^2 = \frac{4\pi^2 a^3}{GM_s} \qquad (6.3)$$

which is just Kepler's third law. Kepler created his laws to describe the motion of the planets in the Solar System, but since the law of gravity is clearly a universal law, other planetary systems must also obey them.

I have made one assumption in this derivation. I have assumed that the Sun is at rest in the centre of the Solar System and that the planets travel around it. This is almost right – the mass of the Sun is much greater than the masses of all the other objects in the Solar System combined – but it is not quite right. As I showed in Chapter 2, the Sun and the planets actually orbit a point called the *centre of mass*, which for the Solar System lies just above the surface of the Sun. It is possible to get the correct answer and still start from the (incorrect) assumption that the planet is orbiting around a stationary object if one assumes the planet is orbiting around an object at the centre of mass with a mass equal to $M_p + M_s$ and replaces the mass of the planet in the equations by the so-called *reduced mass*:

$$\mu = \frac{M_p M_s}{M_p + M_s} \tag{6.4}$$

If we repeat the derivation of Equation 6.3, it changes slightly to become:

$$P^2 = \frac{4\pi^2 a^3}{G(M_p + M_s)} \tag{6.5}$$

with a now being the distance from the centre of mass.

There is one other assumption I have made. I have only considered the gravitational force between a planet and the Sun and assumed the gravitational forces between the planets are negligible. If one wants to do anything more complex than deriving Kepler's laws, such as determining whether the orbit of an object is stable or predicting the position of the Earth millions of years in the future, it is necessary to also take account of the gravitational forces between the planets. There is a problem with this, however. If there are only two objects in a planetary system, it is possible to determine their orbits by solving a few equations with pen and paper. But if there are more than two objects, although the equations are still very simple – the planets' orbits are still governed only by Newton's laws of motion and law of gravity – it is usually only possible to solve these numerically with a computer. The dynamics of planetary systems is a complex subject and much of it is well beyond the scope of this book. However, here are a few key results about the dynamics of planetary systems with only sketchy details of how these are obtained (if you are interested in finding out more, you will have to look up an advanced textbook).

6.2 Stable and unstable orbits

The orbits of most of the objects in the Solar System are stable for the obvious reason that if their orbits were unstable the objects would not stay there very long. Computer simulations have shown the reason why the regions between most of the planets are empty is that the gravitational fields of the planets make orbits in these regions unstable. The obvious exceptions are the regions between the orbits of Mars and Jupiter and outside the orbit of Neptune, where it is possible for a small object to orbit the Sun in a stable orbit. The existence of large numbers of objects in the asteroid and Edgeworth–Kuiper belts (Chapter 7) suggests that the Solar System is in some sense 'full to capacity' because all the regions in which stable orbits are possible contain many objects. One group of objects that do have unstable orbits are the Centaurs, asteroids with very eccentric orbits that cross the orbits of Saturn, Uranus and Neptune; within a 100 million years or so, a Centaur will travel too close to one of the large planets and most likely be ejected from the Solar System.

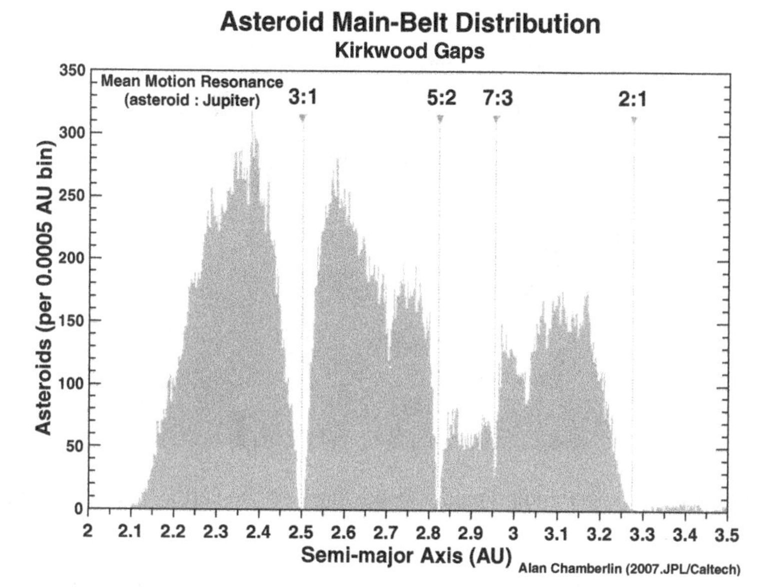

Figure 6.2 The distribution of the semi-major axes in the asteroid belt. The vertical lines show the Kirkwood gaps, where an asteroid's orbital period and Jupiter's orbital period have a simple numerical ratio. (courtesy: NASA/JPL).

Even within the asteroid belt, not every possible orbit is stable. Figure 6.2 shows the semi-major axes of the orbits of the asteroids. There are a number of gaps in the distribution where there are very few asteroids, which are called the *Kirkwood gaps* after the American astronomer, Daniel Kirkwood, who discovered them in 1867. Asteroids in these gaps have orbital periods equal to 1/4, 1/3, 2/5, 3/7 or 1/2 that of Jupiter, with the result that they come close to Jupiter more often than other asteroids, enhancing Jupiter's perturbing effect on their orbits. These are called *orbital resonances* by analogy with the resonant frequency of a musical instrument, the natural frequency at which the instrument's response is greatest. These are unstable orbital resonances because the effect of Jupiter's gravitational field is to clear these regions of asteroids. There are also groups of asteroids outside the main belt which have semi-major axes equal to 3.96 AU and 4.24 AU, which means their orbital periods are exactly 2/3 and 3/4 that of Jupiter. These are stable orbital resonances, because the effect of Jupiter's gravitational field in these orbital resonances is to herd asteroids towards these regions. The most famous object in a stable 2/3 orbital resonance is the former planet, Pluto, which orbits the Sun twice in the time it take Neptune to orbit the Sun three times.

There are also some asteroids with the same semi-major axis as Jupiter. Predating computers by two centuries, the Italian–French astronomer Joseph Lagrange used the traditional pen and paper to solve the so-called 'restricted three-body problem',

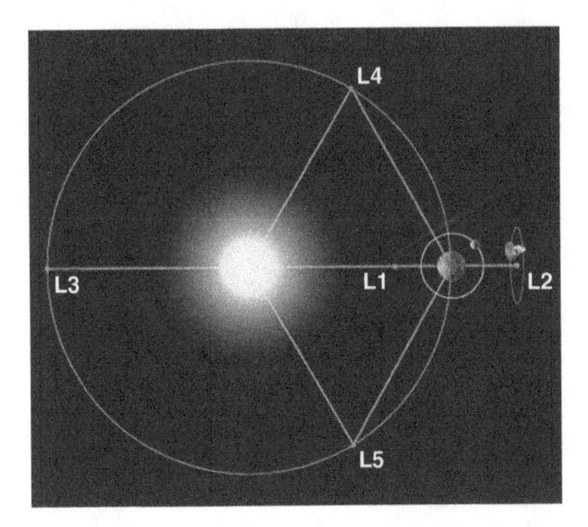

Figure 6.3 The Lagrangian points of the Sun and the Earth. (courtesy: NASA)

in which there are two large objects in a planetary system and one much smaller object. He showed that there are five special points at which the small object could orbit one of the others and experience no net force moving it away from this orbit (Figure 6.3). At three of these *Lagrangian points* – L_1, L_2 and L_3 – the objects lie along a straight line. These are unstable orbits, because if for any reason the small object moves away from the exact Lagrangian point it will gradually drift further away. The other two Lagrangian points – L_4 and L_5 – are stable, however, and once an object is at one of these points it will stay there. A small object at either L_4 or L_5 forms an equilateral triangle with the two large objects. The Trojan asteroids are at the Lagrangian points of Jupiter and the Sun, and thus have the same semi-major axis as Jupiter but travel around the Sun 60° behind or ahead of the planet.

Some comets pass through the Solar System once and are then never seen again (Chapter 7). An object will escape from the Sun's gravitational field if its kinetic energy is greater than its gravitational potential energy. Its escape velocity is therefore given by

$$\frac{v_{\text{esc}}^2}{2} = \frac{GM_s}{r} \tag{6.6}$$

in which r is the object's distance from the Sun. The real situations in which an object might escape from the Solar System include the gravitational interaction between a giant planet and an asteroid, which might give the asteroid a velocity greater than the Sun's escape velocity, and the collision of two asteroids, which might produce fragments with a velocity above the escape velocity. As long as the object's velocity is greater than the escape velocity, it does not matter in which direction it is travelling – unless it collides with the Sun or a planet, it will eventually escape from the Solar System. The path followed by an object that is not gravitationally bound to the Sun is a hyperbola, another member of the family of mathematical curves of which the ellipse is a member (Appendix 2).

6.3 Tidal forces

The dynamics of the Solar System are completely governed by a very simple physical law, Newton's law of gravity, but it is remarkable how such a simple law can give rise to such a diverse range of phenomena. Tidal forces occur because the objects in the Solar System are not points, and so the gravitational forces on two sides of an object are not exactly the same. This is shown pictorially in Figure 6.4, in which I have drawn vectors showing the directions and magnitudes of the gravitational forces exerted by a large object at four positions on a small object. The vectors all point towards the centre of the large object because the gravitational field of a sphere is spherically symmetric. First, consider the forces acting on the bottom and top (as you look at the figure) of the small object. These forces have slightly different directions, and both have a component that is acting towards the centre of the small object, with the result that this object is being squashed in the vertical direction. Now consider the other two vectors. These point in the same direction, but one is obviously bigger than the other because the gravitational force on the near side of the small object is bigger than the gravitational force on the far side. The consequence of this, although it is perhaps less obvious, is that the small object is effectively being stretched in the horizontal direction. The small object is producing a similar, although smaller, tidal effect on the large object.

There are two very visible consequences of tidal forces. The first, of course, is the ocean's tides, which are a consequence of the Moon's tidal forces on the Earth. The best way to understand these is to imagine that the ocean covers the Earth completely except for one small spot of land on which you are standing (Figure 6.5). Because of the difference between the Moon's gravitational forces on different parts of the Earth, the Earth is effectively being stretched and compressed. Because the

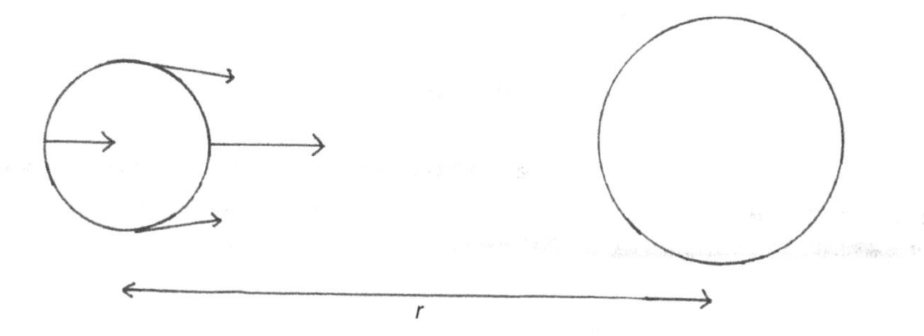

Figure 6.4 The gravitational forces exerted by a large body at four points on a small body. The vectors show the magnitudes and directions of the forces.

These tidal forces are what makes earth not perfectly round.

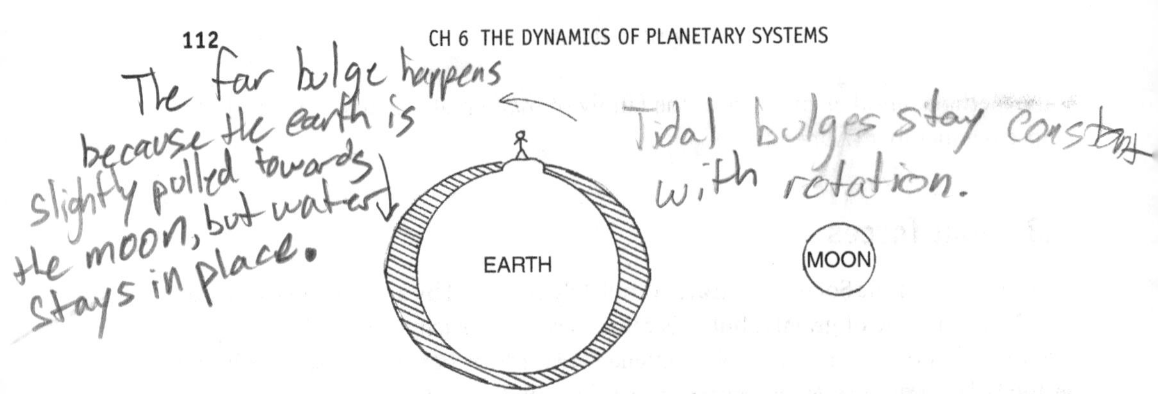

Figure 6.5 The oceans' tides. The hatched area represents the Earth's oceans, and to dramatize the effect of the tides I have shown only one small piece of land.

solid body of the Earth resists these forces better, their effect is greatest on the oceans. The result is that the ocean is deepest at points facing towards and away from the Moon (I have exaggerated the size of these *tidal bulges* in the figure because if drawn to scale they would be too small to see). Now imagine that you are standing on the Earth's one island. As the Earth rotates, you will see the surface of the ocean rise and fall every 12 hours.

It is fairly easy to see where in the Solar System tidal forces are likely to be important. Let us suppose that in Figure 6.4 there is a 1-kg test mass at the centre of the small object. The gravitational force produced by the large object on the test mass is $F = GM/r^2$, in which M is the large object's mass. The gravitational force is slightly less if the test mass is on the nearside of the small object and slightly greater if the test mass is on the far side. The difference between the gravitational forces on the near and far sides is given by a simple piece of calculus:

$$\delta F \simeq \frac{dF}{dr}\delta r \rightarrow \delta F \simeq \frac{2dGM}{r^3} \tag{6.7}$$

in which d is the small object's diameter. This is the tidal force. The equation shows where tidal forces will be important. The dependence on mass and the strong inverse dependence on distance – stronger than the r^2 dependence in Newton's law of gravity – show that tidal phenomena are likely to be most important close to large objects. In the Solar System, therefore, the two places where one would expect to see interesting tidal effects are close to the Sun and Jupiter. The Sun is so massive that it even has a significant tidal effect at the distance of the Earth. By using Equation 6.7, it is possible to show that the Sun's tidal effect on the Earth is roughly half that of the Moon's. The tidal effect of the Sun is the explanation of why the oceans' tides are greatest at new and full Moon, when the Sun, Earth and Moon are in an approximate line.

The other visible consequence of tidal forces is even more obvious than the ocean's tides. As the Earth rotates, it is alternatively stretched and compressed by the Moon's tidal forces, and this tidal pummelling generates heat in the same way that a tennis ball gets hot if it is repeatedly squeezed. This energy has to come from somewhere, and the only possible source is the rotational energy of the Earth. Therefore, with no equations at all, we can predict that the Earth must be slowing down. There is plenty of evidence this is true. The growth bands of fossil bivalve shells and corals imply that 350 million years ago the year was approximately 400 days long; records of eclipses show that the day has lengthened slightly over the last 2000 years; and precise measurements with atomic clocks even show this deceleration from day to day. The Earth's tidal force on the Moon is much greater than the Moon's tidal force on the Earth, and the Moon has already slowed down so much that it is now rotating on its axis only once every 27.3 days, the same time that it takes to travel once round the Earth. The equality of the Moon's orbital and rotational periods is no coincidence because once an object is in *synchronous rotation* it does not slow down any more; no more energy is lost because the tidal bulges are effectively frozen in position, in the same way that no heat is produced if a tennis ball is not repeatedly squeezed but just squeezed once and not released. The consequence of this synchronous rotation is that the Moon always shows us the same face, and so the most visible consequence of tidal forces is the 'Man in the Moon'.

We can see another interesting tidal effect by considering the Earth–Moon system as a whole. As the rotation of the Earth slows down, the angular momentum of the Earth–Moon system must remain a constant because of the law of conservation of angular momentum. The angular momentum of this system has two main components: the rotational angular momentum of the Earth and the orbital angular momentum of the Moon (the rotational angular momentum of the Moon is now negligible). The angular momentum of the Earth–Moon system is therefore

$$J = \frac{2}{5} M_e R_e^2 \omega_e + \mu_m d^2 \Omega_m \tag{6.8}$$

in which M_e, R_e and ω_e are the mass, radius and rotational angular velocity of the Earth; μ_m and Ω_m are the reduced mass and orbital angular velocity of the Moon; and d is the distance from the Moon to the Earth. I have made the approximation that the moment of inertia of the Earth is $(2/5)M_e R_e^2$, which is not quite true (Chapter 4) but good enough for my purpose here. Equation 6.5 gives an expression for Ω_m:

$$\Omega_m^2 = \frac{G(M_e + M_m)}{d^3} \tag{6.9}$$

Using this equation, we can rewrite Equation 6.8 as

$$J = \frac{2}{5} M_e R_e^2 \omega_e + \mu_m \sqrt{dG(M_e + M_m)} \tag{6.10}$$

J is a constant because of the law of conservation of angular momentum. The first term on the right-hand side is decreasing because the rotation of the Earth is slowing down, and so the second term must be increasing in order that the total angular momentum remains constant. The only way this can happen is if d, the distance between the Earth and the Moon, is gradually increasing. This has the interesting consequence that when the Earth itself is eventually in synchronous rotation, the Moon will look much smaller because it will be further away (it will also only be visible from one hemisphere).

There are many other objects in the Solar System in synchronous rotation, including many of the moons and the Pluto-Charon system. Both Pluto and Charon have slowed down so much that they now show the same face to the other object – the future of the Earth–Moon system. An interesting exception is the planet Mercury. Mercury's orbit is quite eccentric, which means that the Sun's tidal force varies as the planet moves around the Sun. The consequence is that the planet is locked in a 3 : 2 *spin–orbit resonance*, in which the planet rotates three times in the time it takes to travel around the Sun twice.

Many other phenomena in the Solar System are, when one looks closely, also the result of tidal forces. One of the most interesting groups of objects in the Solar System are the four giant moons of Jupiter discovered by Galileo in 1609 when he pointed one of the first telescopes at the planet. Io, Europa, Ganymede and Callisto are all moons of the same planet and their masses are not very different, but they are the most spectacularly diverse group of objects in the Solar System (Figure 1.4). Io, the closest moon to Jupiter, must be the strangest object in our planetary system. Its bizarre colours – oranges, reds, whites and blacks – led one of the scientists who saw the first images of the moon from Voyager 2 to claim it looked like a pizza. The Voyager scientists discovered that Io has more volcanoes per square kilometre than any other world in the Solar System. The volcanoes and lurid colours are connected. The volcanoes belch out sulfur-rich compounds, which then freeze and fall back as snow onto the moon's surface. Sulfur and chemical compounds containing sulfur have vivid, if not very tasteful, colours, and it is this layer of snow, many metres thick, which is responsible for the moon's bizarre appearance.

The next moon out, Europa, is completely different. The Voyager images showed that it has a smooth, shiny surface covered by a network of fine lines. The NASA scientists realized that the absence of the usual topography – hills, valleys and craters – means the moon must be covered by a thick layer of ice. The fine lines are cracks in the ice, and the scientists speculated that there might be an ocean under the ice and that water might flow up through the cracks and fill in any new craters (there are almost no craters on the surface of Europa at all). Twenty years later, the Galileo spacecraft found new evidence from the properties of Europa's magnetic field for the existence of this ocean. Because water is one of the basic requirements for life (at least as we know it), Europa's hidden ocean has now risen close to the top of the list of places to look for extraterrestrial life (Chapter 9).

The third and fourth moons, Ganymede and Callisto, are also unique worlds but in a more subdued way. The third moon, Ganymede, has strange grooves across its surface and fewer craters than our moon (Figure 6.6), which suggests its surface is younger. Callisto, the outermost large moon, has a dark surface and, to compensate

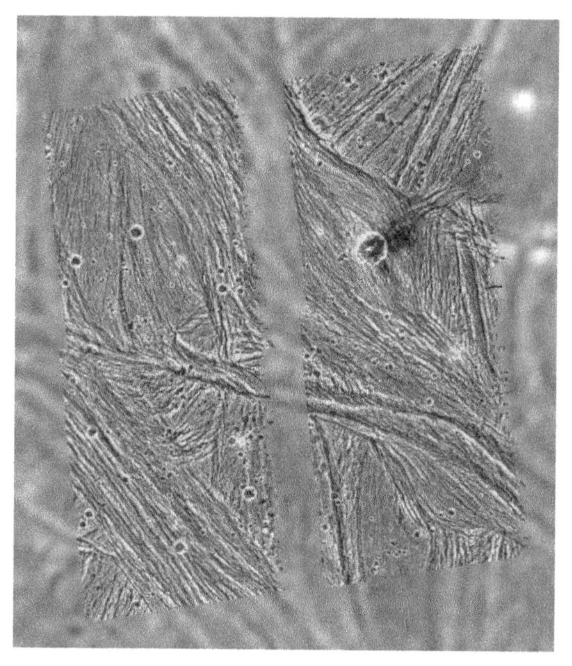

Figure 6.6 Image of the strange grooved terrain on Ganymede. The detailed images were taken by the Galileo spacecraft in 1996 and are superimposed on the lower resolution image taken by Voyager 2 in 1979 (courtesy: NASA).

for the lack of them on Europa, is so densely covered in craters that it may have the oldest surface of any large world in the Solar System.

The explanation of this spectacular diversity is tidal forces. Jupiter's tidal forces are strongest on Io, and it is the tidal heating of the moon's interior that is the natural explanation of the spectacular volcanic activity and hence the lurid colours.[1] Although the tidal forces are not so extreme on the second moon, Europa, they are still strong enough for tidal heating of the moon's interior to be important, which is one of the reasons why scientists think there might be an ocean under the ice. The probable explanation of the cracks in the surface and the lack of any craters is the tidal flexing that occurs as the moon travels around the planet; water flows up through the cracks and over the surface, filling up any new craters and then freezing. The tidal forces are less severe on Ganymede, although they may explain the strange grooves, and they are very weak indeed on Callisto, which is why it has an old dead surface. Thus the way Jupiter's tidal force decreases with distance from the planet is a simple explanation of what seems, at first sight, an inexplicable group of objects.

[1] The moon is now in synchronous rotation with Jupiter but its eccentric orbit means that it is still stretched and compressed as it travels around the planet.

A final tidal phenomenon is responsible for one of the prettiest sights in the Solar System: the rings of the giant planets. If a moon is too close to a planet, it will be ripped apart by the planet's tidal forces. We can make a rough estimate of the distance at which this happens using the basic equation for tidal forces (Equation 6.7). Consider the forces on a test mass of 1 kg on the far side of the moon from the planet. The tidal force on this test mass is given by the difference between the planet's gravitational force at this point and the planet's gravitational force at the centre of the moon:

$$F_{\mathrm{tidal}} \simeq \frac{2GM_{\mathrm{p}}R_{\mathrm{m}}}{r^3} \tag{6.11}$$

in which M_{p} is the mass of the planet, R_{m} is the radius of the moon and r is the distance between the planet and the moon. Let us suppose that the moon is at the critical distance at which the disruptive tidal forces are just balanced by the gravitational forces holding the moon together. The tidal force on the test mass must equal the gravitational force of the moon on the test mass:

$$\frac{GM_{\mathrm{m}}}{R_{\mathrm{m}}^2} = \frac{2GM_{\mathrm{p}}R_{\mathrm{m}}}{r^3} \tag{6.12}$$

If we replace the mass of the planet by its volume times its density, ρ_{p}, do the same for the moon (density ρ_{m}), and do a little rearranging, we get the following equation for the critical distance:

$$r_{\mathrm{crit}} = 2^{\frac{1}{3}} \left(\frac{\rho_{\mathrm{p}}}{\rho_{\mathrm{m}}}\right)^{\frac{1}{3}} R_{\mathrm{p}} \tag{6.13}$$

In 1848 the French astronomer, Edouard Roche, made a much more detailed analysis of the problem and determined that the critical distance is

$$r_{\mathrm{crit}} = 2.456 \left(\frac{\rho_{\mathrm{p}}}{\rho_{\mathrm{m}}}\right)^{\frac{1}{3}} R_{\mathrm{p}} \tag{6.14}$$

which is now known as the *Roche limit* and is approximately a factor of 2 greater than the value given in Equation 6.13. The planetary rings are all inside the Roche limit and the large moons are all outside the Roche limit, which suggests that tidal forces are the ultimate cause of the rings. It is still unclear, however, whether the rings of Saturn (Figure 1.1), which would surely feature as one of the seven wonders of the Solar System, and the other planetary rings were formed by the tidal disruption of moons that strayed too close to their planets, or whether they are formed of material left over from the formation of the Solar System which was prevented from coalescing into moons by tidal forces.

Exercises

1 A baseball pitcher can throw a fastball at a speed of about 150 km/h. What is the largest asteroid from which he can throw the ball fast enough that it will escape from the asteroid's gravitational field? You should assume that the asteroid is spherical and has a density of 3000 kg m^{-3}.

2 The Moon's tidal forces are gradually slowing down the rotation of the Earth. Calculate how much smaller the Moon will appear in the sky when day on Earth is 36 hours long (Mass of Earth: 6×10^{24} kg; mass of Moon: 7×10^{22} kg; radius of Earth: 6378 km; Earth–Moon distance: 3.8×10^5 km; orbital period of Moon: 27.3 days).

3 In 1994 the comet Shoemaker–Levy 9 crashed into Jupiter. When the comet got to within 9×10^4 km of the planet it broke into bits because of the planet's tidal forces.

(a) On the assumption that the diameter of the comet's nucleus was about 2 km, calculate the approximate tidal force on a unit mass of the comet at the time it broke up (mass of Jupiter: 1.8×10^{27} kg).

(b) If the comet's nucleus was made of solid rock, estimate the force holding it together. (You should ignore gravity and estimate the force across a cross-section of the comet; the tensile strength of rock is 5×10^8 N m^{-2}.)

(c) Estimate the total tidal force acting on the comet by multiplying your answer from (a) by the mass of the comet. Compare your answer with the answer to (b) and comment on its implications. (You may assume the density of the comet's nucleus is approximately 3000 kg m^{-3}.)

Figure 1.1 The eight planets in our planetary system. (a) Mercury (Messenger); Venus (Pioneer Venus Orbiter); Earth (Apollo 8). (b) Mars (Viking Orbiter); Jupiter (Voyager 2); Saturn (Voyager 2). (c) Uranus (Voyager 2); Neptune (Voyager 2) (courtesy: NASA).

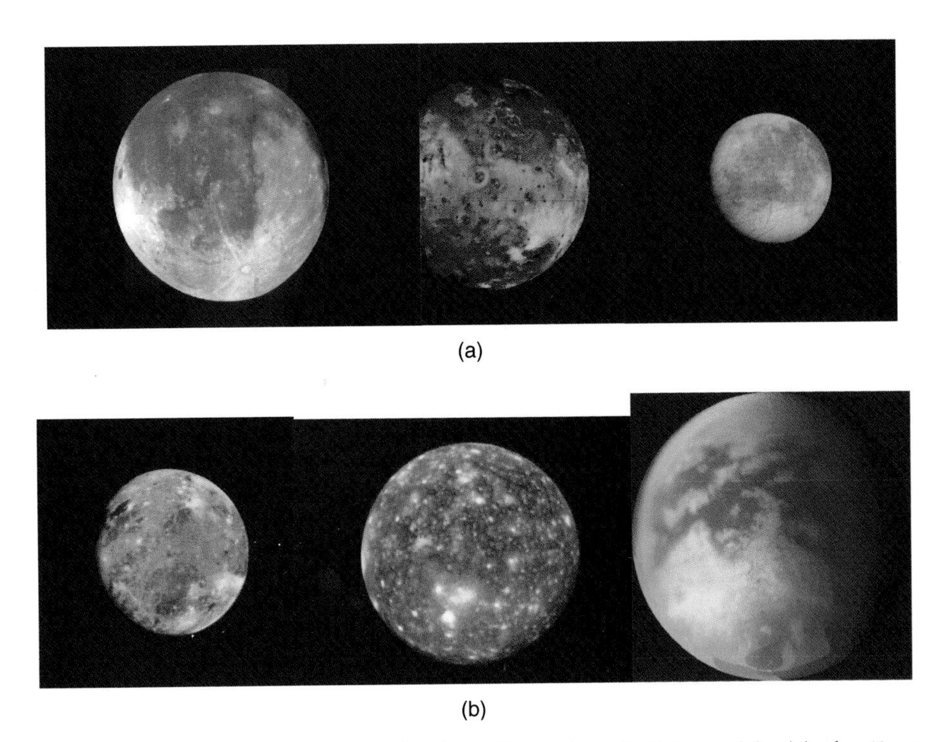

Figure 1.4 The six largest moons in the solar system (not to scale). (a) the Moon (Clementine); Io (Voyager 1); Europa (Voyager 1). (b) Ganymede (Voyager 1); Callisto (Voyager 1); Titan (Cassini) (courtesy: NASA and ESA).

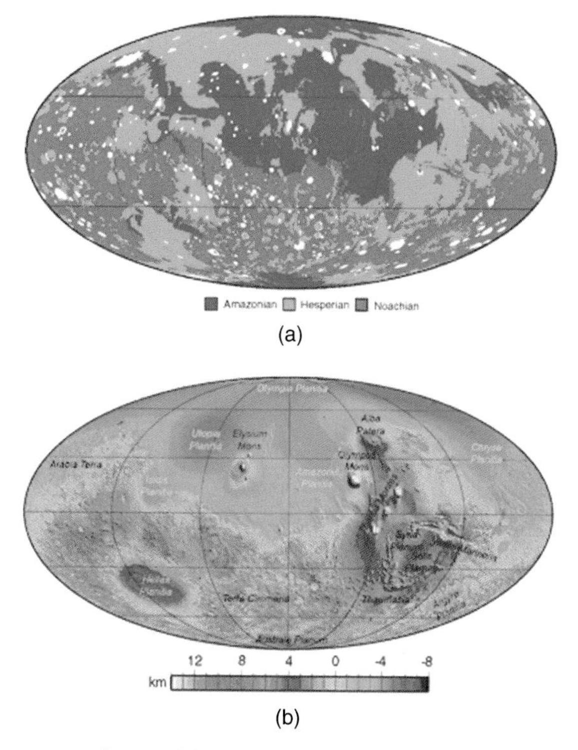

Figure 3.12 Two maps of Mars. (a) is a geological map showing the parts of the surface that were formed during the three epochs of Martian history: Noachian (orange), Hesperian (green), Amazonian (blue). The white areas are where debris from recent large impacts has covered geological structures and earlier craters, making it impossible to estimate the age of the surface beneath. (b) is a topographic map made by the Mars Orbiter Laser Altimeter (see text) (from Solomon *et al.* 2005, *Science*, **307**, 1214 reprinted with permission from AAAS)

Figure 3.13 Topographic map of Venus made by Magellan, in a mercator projection (north is at the top). The lowest regions are shown as blue, the highest as red (courtesy: NASA).

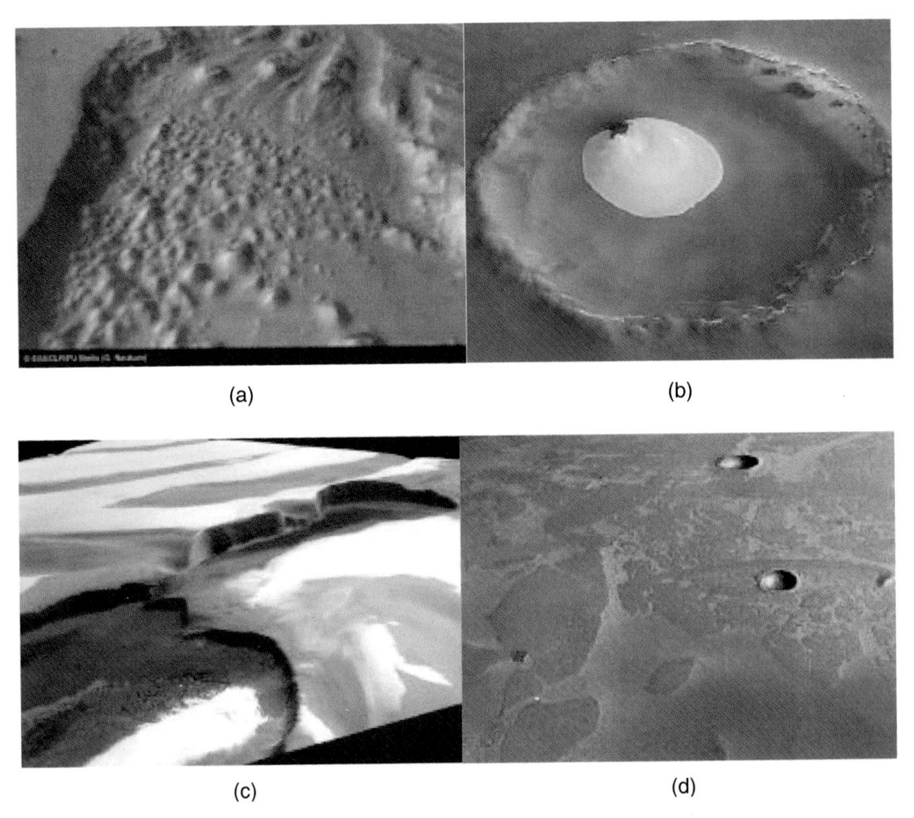

Figure 3.15 Four images from Mars Express that show the presence of water on Mars: (a) an outflow channel; (b) ice in a crater; (c) a mixture of ice and dust at the North Pole; (d) a possible frozen sea covered by dust (courtesy: ESA).

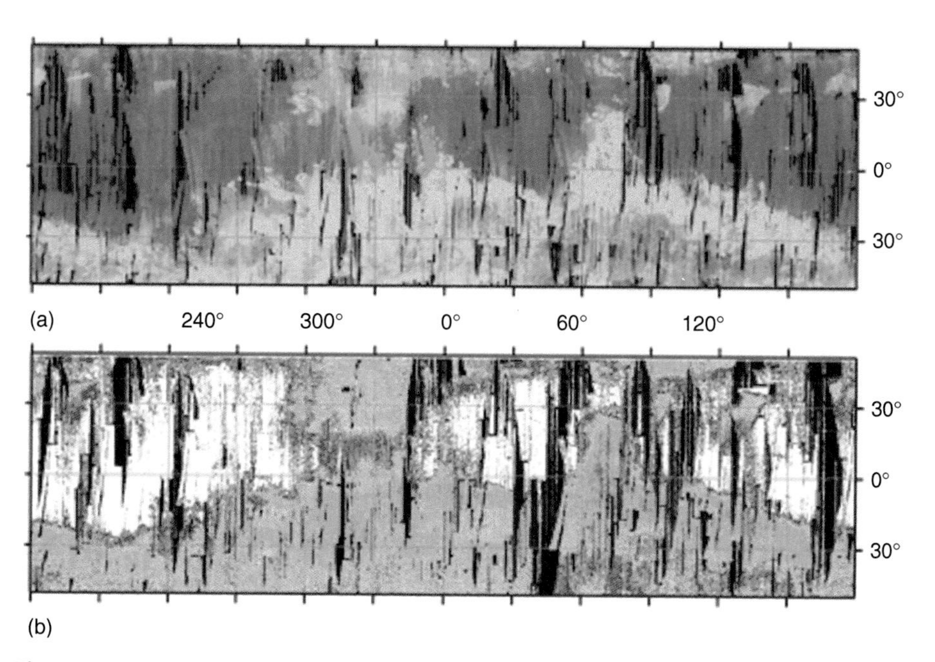

Figure 3.16 Maps made by Mars Express of the distribution of two important minerals (Mercator projection, north at the top). (a) shows the distribution of pyroxene, an important constituent of basalt rocks, with yellow showing its presence and blue where there is none. (b) shows the distribution of iron oxide, with white and red indicating its presence (from Bibring *et al.* 2006, *Science*, **312**, 400 reprinted with permission from AAAS).

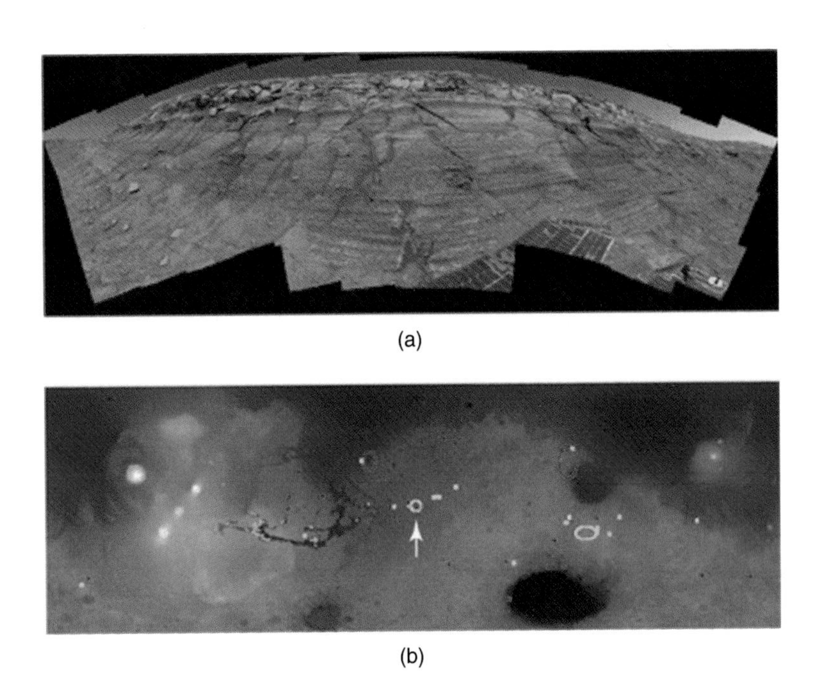

Figure 3.17 A local and global view of the geology of Mars. The local view in (a) is an image taken by the Mars rover Opportunity of a few metres of Burns Cliff, a rock outcrop that is part of the rim of Endurance Crater. The global view in (b) shows a topographic map (lighter colours imply a higher elevation) on which are superimposed the location of several important minerals: sulfates (blue), phyllosilicates (red), other hydrated minerals (yellow). The arrow shows where Opportunity landed (from Bibring *et al.* 2006, *Science*, **312**, 400 reprinted with permission from AAAS).

Figure 5.5 A false-colour image of the surface of Titan made using radar data from Cassini. Areas that reflect radio waves poorly are shown as blue

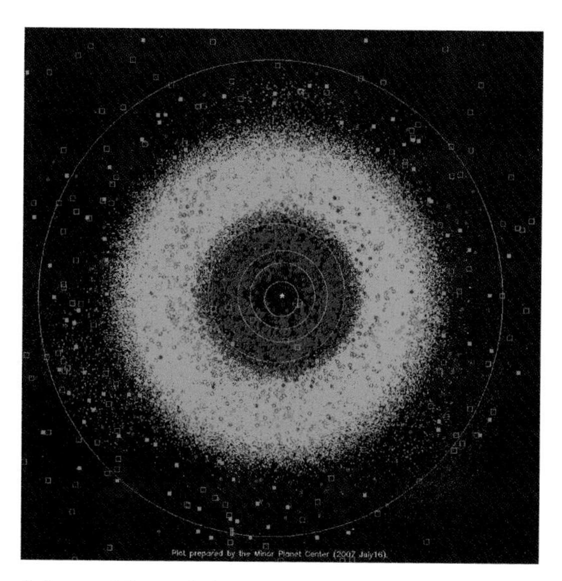

Figure 7.4 Plot of the positions of the comets and asteroids on 16 July 2007, produced by the Minor Planet Center, which has the responsibility of keeping track of all the small objects in the Solar System. The large circles show the orbits of the planets, the outermost circle being that of Jupiter. The squares are comets and the circles and dots are asteroids. The green dots are the asteroids in the main belt; the two clumps of blue dots at the bottom right and left are the Trojan asteroids (Chapter 6); the red dots are Near Earth Objects (reproduced courtesy of Gareth Williams, Minor Planet Center).

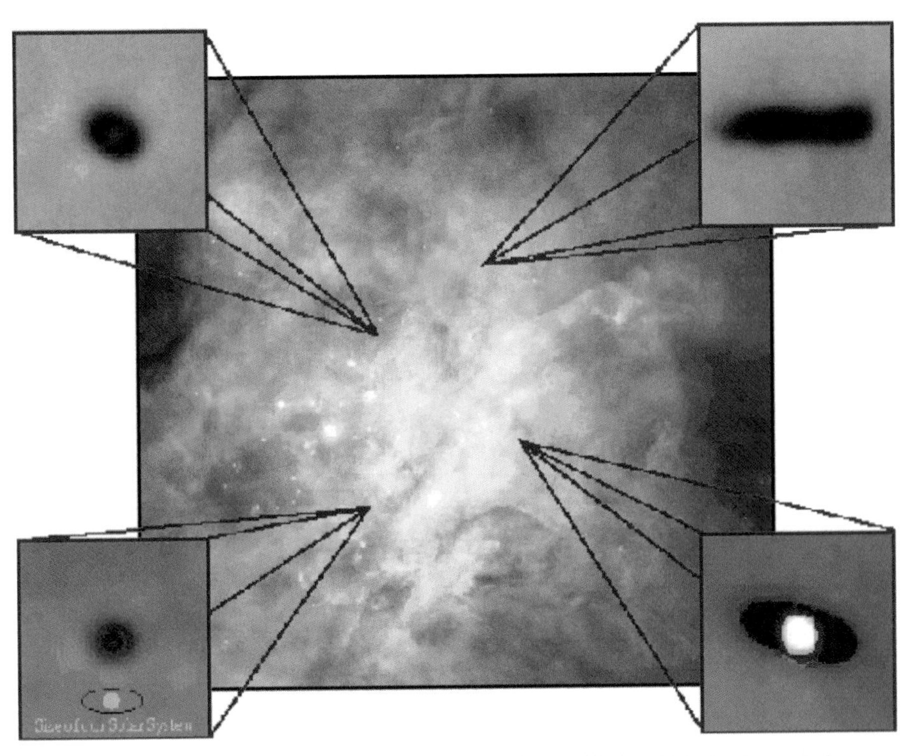

Figure 8.1 Image of part of the Orion Nebula taken with the Hubble Space Telescope. The insets show the silhouettes of dusty discs against the light of the nebula (reproduced courtesy of C.R. O'Dell/Rice University and NASA).

7 The small objects in planetary systems

Gentlemen, I would rather believe two Yankee professors would lie than believe that stones fall from heaven

Thomas Jefferson

7.1 The evidence of the meteorites

The big objects in planetary systems are interesting because of the many geological, atmospheric and biological processes that still occur on them – and also of course because we live on one. The small objects in planetary systems are interesting for an entirely different reason. These objects have almost certainly not changed much in the last 4.5 billion years, and their properties therefore constitute a book, if only we can read it, about the origin of the Solar System. There are two reasons for thinking this, one theoretical and one empirical.

The theoretical one is that small objects cool faster than big objects, for the same dimensional reason that elephants have thicker legs than mice and that the largest birds are still small compared to the largest land animals. The luminosity of an object is proportional to its surface area (R^2), whereas its energy content is proportional to its volume (R^3), which means the rate at which it loses energy is proportional to $1/R$. The small objects in the Solar System were probably hot when first formed (Chapter 4), but they would soon have lost this initial heat. Since it is the Earth's internal heat that is responsible for the interesting geological processes we now see on its surface, these small objects, the asteroids and comets, have from a geologist's point of view been boring places for a very long time. They have been

Planets and Planetary Systems Stephen Eales

in deep freeze since the beginning of the Solar System, and so their properties may tell us something about how this happened.

The empirical reason is the ages of meteorites. Thomas Jefferson, quoted above, was probably quite right to be sceptical about the connection between the streams of light seen in the sky (meteors) and the strange rocks found on the ground (meteorites) given some of the other strange scientific ideas that were in vogue at the time. The most direct evidence that meteorites do fall from the sky are the 10 000 meteorites that have been found lying on the ice in Antarctica, because it is hard to think of how else they could have got there. Scientists divide meteorites into three classes: *irons*, which are composed mostly of iron and other metals; *stones*, which are non-metallic; and *stony-irons*, which are half and half (Figure 7.1). The irons are the ones mostly seen in museums, because they are the easiest to spot on the ground among all the other rocks and because they are the nicest to look at. But in Antarctica, where there is no problem of confusion with other rocks, only about 1 % of the meteorites are irons or stony-irons. The irons and stony-irons must have come from objects in which there has been *chemical differentiation*, which can occur either by heavier elements sinking to the centre of the object, as happened in the Earth, or by different minerals condensing out of a magma at different times. The stones are divided into *achondrites*, in which there is evidence from the elemental abundances that some chemical differentiation has occurred, and *chondrites* in which there is no evidence for this. Of the meteorites found in Antarctica, about 5 % are achondrites, and so 94 % of meteorites are objects in which no chemical differentiation has occurred and which have therefore possibly never been molten.

Meteorites fall from the sky, but what is their ultimate source? There is evidence that about 10 of the Antarctic meteorites come from Mars, based on the abundances of noble gases within the meteorites, which are very similar to those in the Martian atmosphere. There is also a handful of other meteorites that probably come from the Moon. The gravitational field of both objects is sufficiently weak that if one

Figure 7.1 Two meteorites – on the left an iron, on the right a stone (courtesy: NASA/JPL).

of them is hit by a rock, fragments from the collision may escape into space and eventually (although very rarely) land on another planet. The remaining 99.9 % of meteorites are either fragments from collisions within the asteroid belt or the debris left over after all of a comet's volatile material has been boiled away. The evidence for this is, first, that these are the only obvious sources – a simple model implies there have been many collisions between asteroids (see below)– and, second, the fact that most meteorites come from objects in which chemical differentiation has not occurred and which have therefore never been molten or have only been molten for a short time, which implies small objects.

The hard evidence that asteroids and comets are primitive objects, in which not much has happened since the dawn of the Solar System, is the ages of the meteorites. We can estimate these by the powerful technique of radioactive dating, used in subjects as far apart as geology and archaeology. Although the technique's basic idea is very simple, there is a big practical problem that is often skated over in introductory textbooks, and since the way this problem is overcome is actually quite interesting, we will now consider the practical details of how this technique is applied to meteorites.

A few of the radioactive decay sequences that are useful in estimating the ages of meteorites are

$$^{40}\text{K} \rightarrow {}^{40}\text{Ar} \quad \text{Half-life} = 1.25\ \text{Gyr}$$
$$^{87}\text{Rb} \rightarrow {}^{87}\text{Sr} \quad \text{Half-life} = 49\ \text{Gyr}$$
$$^{238}\text{Ur} \rightarrow {}^{206}\text{Pb} \quad \text{Half-life} = 4.47\ \text{Gyr}$$

The basic idea of the technique is to measure the amounts of the 'daughter' isotope, for example^{87}Sr, and the 'parent' isotope, in this case ^{87}Rb. We can then estimate the age of the sample in the following way. Applied to this particular sequence, the fundamental equation of radioactive decay is

$$[^{87}\text{Rb}] = [^{87}Rb]_0\, e^{-\frac{t}{1.44\tau}} \tag{7.1}$$

in which τ is the half-life of the decay sequence, $[^{87}\text{Rb}]$ is the number of Rb atoms present in the sample after a time t, and $[^{87}\text{Rb}]_0$ is the number of atoms that were in the sample at $t = 0$. The number of daughter atoms is simply

$$[^{87}\text{Sr}] = [^{87}\text{Rb}]_0 - [^{87}\text{Rb}] \tag{7.2}$$

and by eliminating $[^{87}\text{Rb}]_0$ between the two equations, we can obtain the following equation:

$$[^{87}\text{Sr}] = [^{87}\text{Rb}](e^{\frac{t}{1.44\tau}} - 1) \tag{7.3}$$

We may assume that we know the half-life of this particular decay from laboratory measurements, and so if we have measured the amount of strontium and rubidium in the sample today, it should be a simple matter to calculate the remaining unknown: the age of the sample.

The practical problem is that in writing down Equation 7.2 I made one huge assumption. I implicitly assumed there was no strontium originally in the sample, which is unlikely to be true. It is still possible to overcome this problem if there is another isotope of strontium *not* formed by radioactive decay. In this case there is: ^{86}Sr. Let us now assume that the ^{87}Sr in the sample today has two components: one present when the mineral was formed and one produced subsequently by radioactive decay:

$$[^{87}\text{Sr}] = [^{87}\text{Sr}]_0 + [^{87}\text{Sr}]_r \tag{7.4}$$

The first term on the right-hand side is the strontium originally in the sample, the second term the strontium produced by radioactive decay since the mineral was formed. Now divide all the terms in this equation by the number of atoms of the other strontium isotope, which is a constant in the sample because it is not radioactive:

$$\frac{[^{87}\text{Sr}]}{[^{86}\text{Sr}]} = \frac{[^{87}\text{Sr}]_r}{[^{86}\text{Sr}]} + \frac{[^{87}\text{Sr}]_0}{[^{86}\text{Sr}]} \tag{7.5}$$

We also have to add one suffix to Equation 7.3 because the strontium in that formula is the strontium produced by radioactive decay:

$$[^{87}\text{Sr}]_r = [^{87}\text{Rb}](e^{\frac{t}{1.44\tau}} - 1) \tag{7.6}$$

By eliminating $[^{87}\text{Sr}]_r$ between Equations 7.5 and 7.6, we obtain

$$\frac{[^{87}\text{Sr}]}{[^{86}\text{Sr}]} = (e^{\frac{t}{1.44\tau}} - 1)\frac{[^{87}\text{Rb}]}{[^{86}\text{Sr}]} + \frac{[^{87}\text{Sr}]_0}{[^{86}\text{Sr}]} \tag{7.7}$$

This equation is simpler than it looks. It has the same form as the standard equation for a straight line, $y = mx + c$, if we make the identifications:

$$y = \frac{[^{87}\text{Sr}]}{[^{86}\text{Sr}]}, \qquad m = (e^{\frac{t}{1.44\tau}} - 1), \qquad x = \frac{[^{87}\text{Rb}]}{[^{86}\text{Sr}]}, \qquad c = \frac{[^{87}\text{Sr}]_0}{[^{86}\text{Sr}]}$$

Now consider carefully each of these terms. We can measure x and y for any mineral within the meteorite. Their values will depend both on the amount of radioactive decay that has occurred and on the chemical ratio of rubidium to strontium in the mineral. The other two variables are unknowns. We don't know m because we don't know the age of the rock and we don't know c because we don't know the original ratio of the two strontium isotopes. If we plot the pairs of x, y measurements for the different minerals within the meteorite, we will obtain a plot like Figure 7.2, which was actually obtained for a meteorite that fell at Tieschitz in the Czech Republic in 1878. The gradient of this line is m, and so if we measure the gradient and know the half-life of the decay sequence, we can estimate the age of the rock. The intercept of the line on the y-axis is c, and so by measuring this we can estimate the ratio of the two strontium isotopes present in the rock when it was formed.

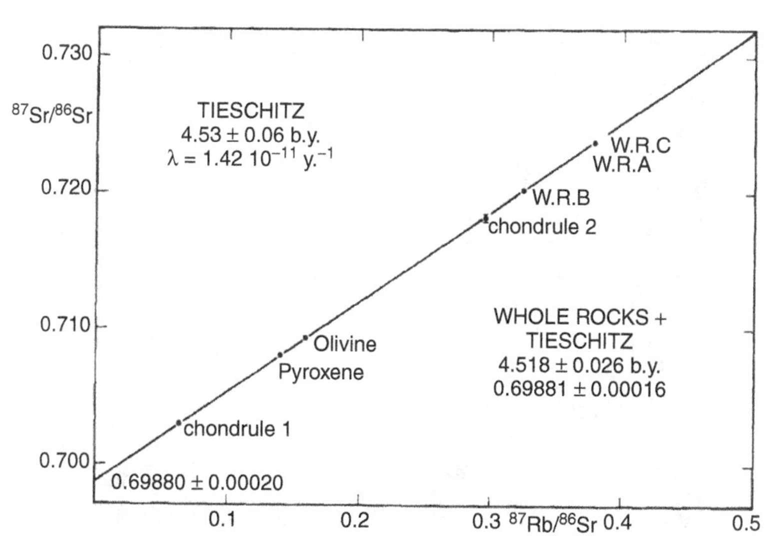

Figure 7.2 Plot of the $^{87}Sr/^{86}Sr$ and $^{87}Rb/^{86}Sr$ ratios for separate minerals within the Tieschitz meteorite (reproduced from Elsevier from Minster, J.F. and Allegre, C.J. (1979), *Earth and Planetary Science Letters*, **42**, 333).

The age of this particular meteorite is 4.52 ± 0.03 billion years, and most meteorites have almost exactly the same age, older than the oldest rocks on the Earth and the Moon. This is the hard evidence that the small objects in the Solar System were all formed at the same time, which must be when the Solar System itself was formed.

7.2 The asteroid belt

These small objects may constitute a book about the origin of the Solar System, but we haven't yet read very far through it because no spacecraft has landed on an asteroid, although one spacecraft has now made a landing (of a sort) on a comet (see below). The closest that a spacecraft has come to an asteroid have been the close approaches to the asteroids Ida (Figure 1.5) and Gaspra (Figure 7.3) made by Galileo on its way to Jupiter. Both asteroids look more like potatoes than spheres, which is not surprising because they are much smaller than the critical size at which gravity shapes an object into a sphere (Chapter 1); Gaspra, for example, is only about 19 km along its long axis. The largest asteroid, and the first one to be discovered, Ceres, has a radius of approximately 480 km, which puts it just over the critical limit. An image taken with the Hubble Space Telescope shows that it does indeed appear to be spherical (Figure 1.8), and so it meets the criteria for being a dwarf planet (Chapter 1). In September 2007 NASA launched the *Dawn* mission, which will orbit and study Ceres and Vesta, another large asteroid, using many of the same

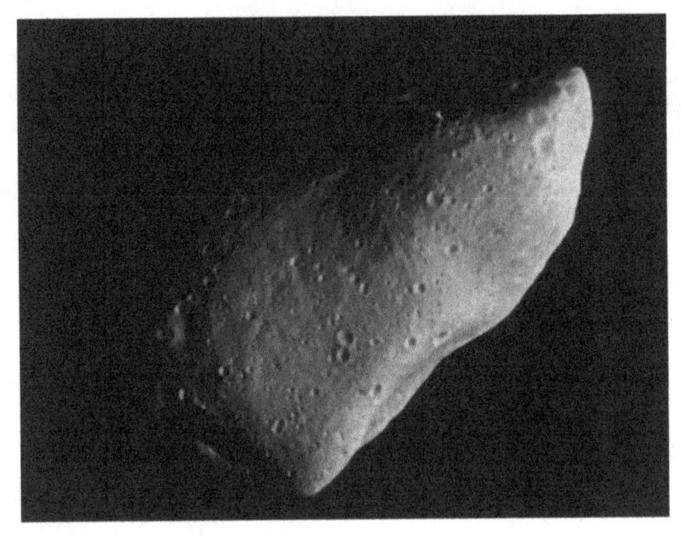

Figure 7.3 Image of the asteroid Gaspra taken by Galileo as it passed the asteroid in 1991 on its way to Jupiter (courtesy: NASA).

techniques that have been used to study planets such as Mars (Chapter 3). Dawn will reach Vesta in 2011 and Ceres in 2015, and so within the next few years we will be able to read much more of this book.

The typical random velocities in the asteroid belt are a few kilometres per second. This is much larger than the typical escape velocity of an asteroid, which means that if two asteroids collide they are likely to be smashed to pieces. A simple theoretical model implies that if enough collisions have occurred the number of asteroids with a radius R should follow a power law,

$$N(R) = N_0 \left(\frac{R}{R_0} \right)^{-\eta} \tag{7.8}$$

with the power-law index, η, having a value of approximately 3.5. This is close to the actual distribution, which is the evidence that there have been many collisions between asteroids. Astronomers have discovered and catalogued approximately 10 000 asteroids, but there must be many more than this because surveys from the Earth can only detect asteroids with radii greater than about 10 km; Equation 7.8 implies there are over 100 000 unknown asteroids with radii between 1 and 10 km. Although most asteroids are small, Equation 7.8 implies that most of the mass in the belt is in a few massive objects. Because most of the mass is in the big asteroids, which *are* in the catalogues, we can estimate fairly reliably the total mass in the belt. This is about 0.05 % of the mass of the Earth, and so all the asteroids together do not add up to one decent-sized planet.

Figure 7.4 shows the positions of all the small objects in the inner Solar System on one day: 16 July 2007. Most of the asteroids have roughly circular orbits between

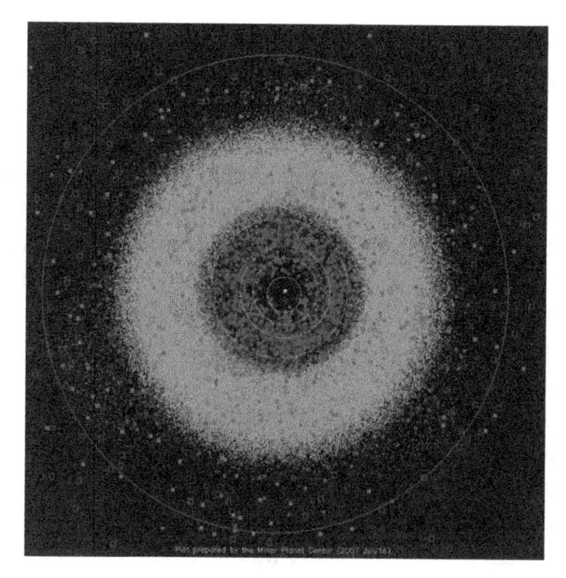

Figure 7.4 Plot of the positions of the comets and asteroids on 16 July 2007, produced by the Minor Planet Center, which has the responsibility of keeping track of all the small objects in the Solar System. A colour reproduction of this figure can be seen in the colour section, located towards the centre of the book. The large circles show the orbits of the planets, the outermost circle being that of Jupiter. The squares are comets and the circles and dots are asteroids. The green dots are the asteroids in the main belt; the two clumps of blue dots at the bottom right and left are the Trojan asteroids (Chapter 6); the red dots are Near Earth Objects (reproduced courtesy of Gareth Williams, Minor Planet Center).

the orbits of Mars and Jupiter. The two groups of blue dots at the bottom of the diagram are the Trojan asteroids, which have stable orbits because they are at the Lagrangian points of Jupiter and the Sun (Chapter 6). The two groups of Trojans and Jupiter form an equilateral triangle, which means that on this day Jupiter must have been close to the top of the diagram.

The red points are asteroids with a perihelion less than 1.3 times the Earth–Sun distance. Many of these actually come within the Earth's orbit. 'Earth-crossing' asteroids and comets are often called *near earth objects* (NEOs). The subject of NEOs is one of the places where astronomy suddenly stops being a beautiful abstract subject and becomes a pressing human concern, because one of these NEOs could potentially hit the Earth. It is not possible to predict precisely how often an NEO should hit the Earth, but a rough calculation suggests that an NEO with a diameter of at least 10 km should hit the Earth every 100 million years or so, which nicely agrees with the typical interval between major extinctions in the fossil record. There is now convincing evidence that the last major extinction, at least, which occurred 65 million years ago and in which roughly half the species on Earth, including the dinosaurs, disappeared, was caused by the impact of an object of about this size (an object with a diameter 10 times larger than this would have sterilized the planet). We may not be able to do much yet if an NEO is heading directly towards the Earth, but various governments have at least taken the first step of spending money on tracking potential NEOs.

The orbit of an NEO is not stable and eventually, as the result of continual changes to its orbit caused by the gravitational fields of the planets and the Sun, it will either collide with the Sun or a planet or more likely be thrown out of the Solar System all together. The source of the NEOs is probably the main belt, because although the orbits there are stable, the gravitational interactions and collisions between the asteroids will occasionally send one onto an orbit that takes it closer to the Sun.

Why is there an asteroid belt at all? The obvious answer is that it is there because Jupiter's huge gravitational field somehow kept a planet from forming in this part of the Solar System. The evidence against this is that all the asteroids together would not have made much of a planet, and one dwarf planet did form here anyway. A more likely answer comes from considering the fact that everywhere we look in the Solar System where there is the potential for stable orbits, we see many small objects, not only in the asteroid belt but in the Edgeworth–Kuiper (EK) belt (see below) and at the Lagrangian points of Jupiter and the Sun. The answer that is probably correct is that when the Solar System was formed there were small objects everywhere, and the small objects that did not have stable orbits were gradually ejected, leaving the ones we see today. This, though, is an integral part of the story of the origin of the Solar System, which we will consider in the next chapter.

7.3 Comets

By tradition, small objects are called *comets* if they have eccentric orbits and grow huge gas tails as they approach the Sun and *asteroids* if they travel in roughly circular orbits and do not have gas tails. Although this conventional distinction is the one we will use here, it is not clear how much meaning it has. An asteroid in the main belt, for example, might have a substantial reservoir of ice and other volatile material, which we never see because the sunlight at this distance from the Sun is too weak to melt it. Images of the nuclei of comets (Figure 7.5) and of asteroids (Figures 1.5 and 7.3) do look very similar. As I will describe in the next chapter, it seems likely that both comets and asteroids are small objects left over from the formation of the Solar System. If there is a distinction between them, it is probably one of geographical origin, with comets being objects that formed further out in the solar nebula, and thus with a higher proportion of volatile material, and asteroids being objects that formed closer to the centre of the nebula; although because of the large-scale migration of small objects that almost certainly occurred after the formation of the planets (Chapter 8) it is unclear whether even this distinction is a genuine one.

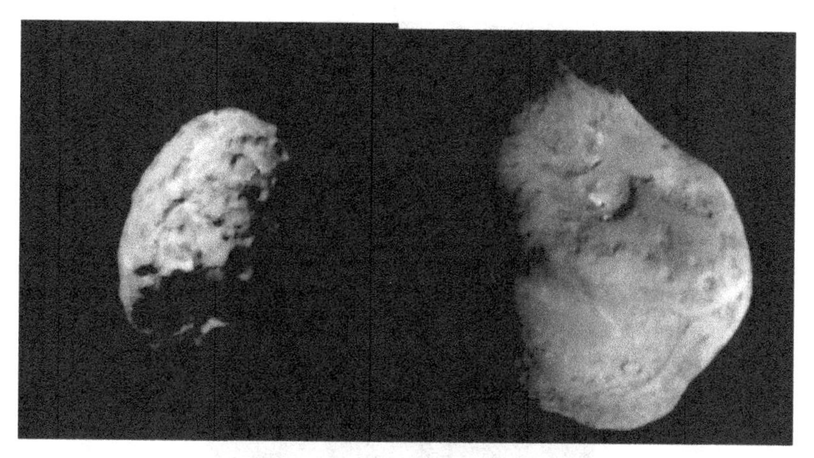

Figure 7.5 Two close-up images of comet nuclei: on the left a picture of Comet Wild 2 taken by Stardust, on the right an image of Comet Tempel 1 taken by the Deep Impact probe 5 minutes before it hit the comet (see below) (courtesy: JPL/NASA).

Again by tradition, a comet is divided into three parts. At a large distance from the Sun, the comet consists only of the solid *nucleus* (Figure 7.5). As it approaches the Sun, some of the volatile material in the nucleus sublimes and gas streams out from the nucleus in all directions to form the *coma*. The flow of gas is strong enough to pull solid chunks of material up to about 1 cm in size from the nucleus, and so the coma is actually a mixture of gas and solid particles (commonly called *dust*). At a large distance from the nucleus, the density of the out-flowing gas and dust is low enough that a *tail* develops, caused by the interaction of the outflowing material with radiation and particles from the Sun. The tail, despite its low density, is the most spectacular part of the comet. There are actually two tails, which are nicely shown in the picture of Comet West in Figure 7.6. The prominent bright tail in the centre is a stream of dust particles swept away from the coma by the pressure of the Sun's radiation. The dust particles are seen in reflected sunlight, and there is a curve in this tail because of the conservation of the angular momentum of the dust particles as they move away from the Sun. The fainter tail seen at the lower right is a stream of ions that is swept away from the coma by the solar wind; this points directly away from the Sun.

There are two ways astronomers have discovered what comets are made of. The first is to make observations of the gas in the coma using telescopes on the Earth. The second is to send a spacecraft out to the comet to study it on the spot or even (see below) to bring comet material back to the Earth.

Spectroscopic observations using telescopes on the Earth are a powerful way of investigating the composition of the coma, but there is the complication that the chemical species seen in the coma are not necessarily the same as those contained

Figure 7.6 Image of Comet West (courtesy: John Laborde).

in the nucleus. For example, atomic hydrogen and oxygen and the OH radical are all seen in comets' comas, but these are produced by the photodissociation of water molecules by sunlight after the water molecules have left the shelter of the nucleus. The chemical composition of the coma will also change with time because of the increase in the intensity of the sunlight as the comet approaches the Sun, causing chemical compounds with higher melting points to sublime from the nucleus. There are also of course the usual intricate technical problems of converting the brightness of spectral lines into the abundances of chemical compounds and elements, but these we will pass over.

Despite these difficulties, astronomers have succeeded in measuring the rates at which different chemical species are produced from the nuclei of several comets. Figure 7.7 shows the results for Comet Lee. As one might expect, the production rates of all chemical compounds increased as the comet moved towards the Sun and then decreased as it receded from the Sun. The figure shows that by far the most abundant chemical species found in the coma of Comet Lee was the OH radical, which is produced by the photodissociation of water molecules. The large abundances of water-derived species in the comas of comets show that water is the most important volatile substance in comets, but the figure shows that Comet Lee also contained many other interesting compounds, including organic ones. The existence of frozen water and other compounds with low melting points suggests that comets must have formed a long way from the Sun.

The other (and much more expensive) way of investigating the composition of a comet is to send a spacecraft. In January 2004, the NASA spacecraft Stardust flew

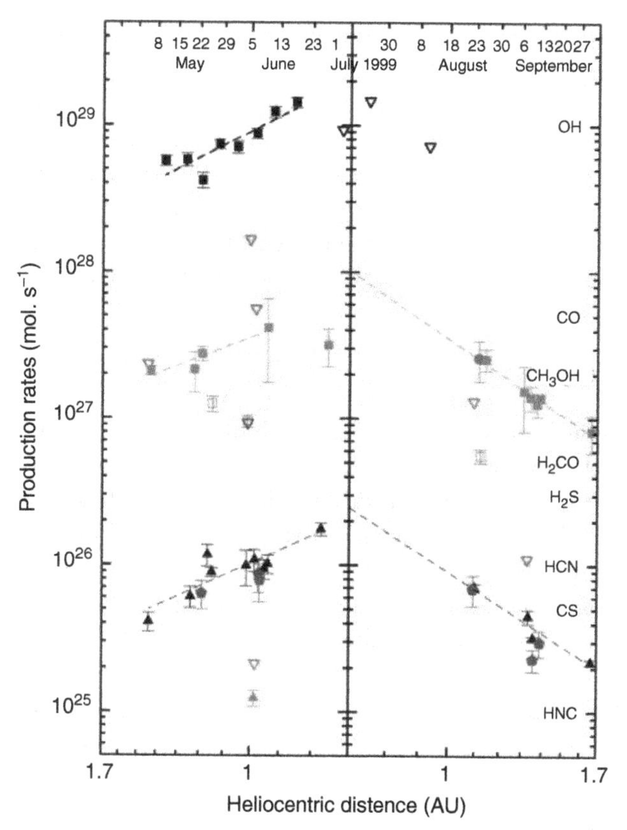

Figure 7.7 The production rates of different molecules measured for Comet Lee. The left-hand side of the figure shows how the production rates of different chemical species changed as the comet approached the Sun, the right-hand side how they changed as it moved away from the Sun (Biver, N., Bockelee-Morvan, D., Crovisier, J. *et al.* (2000), *Astronomical Journal*, **120**, 1554).

past Comet Wild 2 (Figure 7.5), collected solid particles from the coma and brought them back to the Earth. The analysis of this material produced a surprising result. Astronomers had expected this solid material would have a similar mineralogical composition to interstellar dust, because they assumed that comets must have formed a long way from the Sun, where the temperature was never high enough to modify the dust originally in the solar nebula (Chapter 8). The material from the comet, however, contained crystals and minerals that could only have been formed at high temperatures. This tantalizing result must have great significance for the formation of the Solar System, although exactly what its significance is, nobody really knows. Its general implication is that some of the material in comets must have formed close to the Sun, but other material – the ices – must have formed far from the Sun where the temperature was much colder, and so there must have been much more movement of solid material within the original solar nebula than astronomers have usually assumed.

Until the mid-1980s, nobody knew what the nucleus of a comet looked like, because the nuclei of comets were too small to study from the Earth. It is still possible, however, to estimate the size of an object that is unresolved from the Earth as long as it does not shine by its own light. If a comet is a long way from the Sun, so that no gas has sublimed from the nucleus, its brightness depends only on the amount of sunlight it reflects. The flux of the comet at a frequency ν as measured on the Earth is given by

$$F_\nu = \frac{L_\nu A_\nu \pi r^2}{16\pi^2 D^2 \Delta^2} \tag{7.9}$$

in which L_ν is the luminosity of the Sun at this frequency, A_ν is the monochromatic albedo, r is the radius of the nucleus, and D and Δ are the distances of the comet from the Sun and the Earth, respectively (this follows from Equation 1.2 and the fact that a comet's brightness depends inversely on the square of its distance from the Earth). On the assumption that we know the position of the comet and have measured its brightness, there are only two unknowns in the equation: the albedo and the size. We could estimate the size by assuming a typical albedo, but it is possible to estimate both quantities by using an additional piece of information. The sunlight that is not reflected is absorbed by the surface. The absorbed energy is balanced by the energy radiated by the surface in the infrared:

$$L_{\mathrm{IR}} = 4\pi r^2 \epsilon T^4 \tag{7.10}$$

in which ε is the emissivity of the comet in the infrared and T is its temperature. It looks as if I have just introduced more unknowns, but the emissivity in the infrared is approximately 1 and the temperature can be estimated from the frequency at which the infrared radiation is at a maximum, which is given by Wien's law:

$$\nu_{\mathrm{max}} = 5.88 \times 10^{10} T \tag{7.11}$$

in which the frequency is measured in Hz and the temperature in Kelvin. By observing both the optical and infrared radiation from a comet, it is therefore possible to estimate both the size of the nucleus and its albedo. Comets' nuclei are very small, mostly between 1 and 20 km in diameter. They are also surprisingly dark. The albedos of most comets are between 2 and 5 %, making these the darkest objects in the Solar System.

The first time anyone saw the nucleus of a comet was in 1986 when the spacecraft Giotto approached to within 600 km of Comet Halley, which was when the comet last came close to the Sun (Figure 7.8). The Giotto images revealed that the surface of the nucleus is extremely dark, suggesting that the ice in the comet is hidden by a layer of rock or dust. The Giotto observations also showed that gas does not stream uniformly from the surface, but rather that there are jets of gas from certain active spots on the surface. The dark surface and the jets can be explained rather simply.

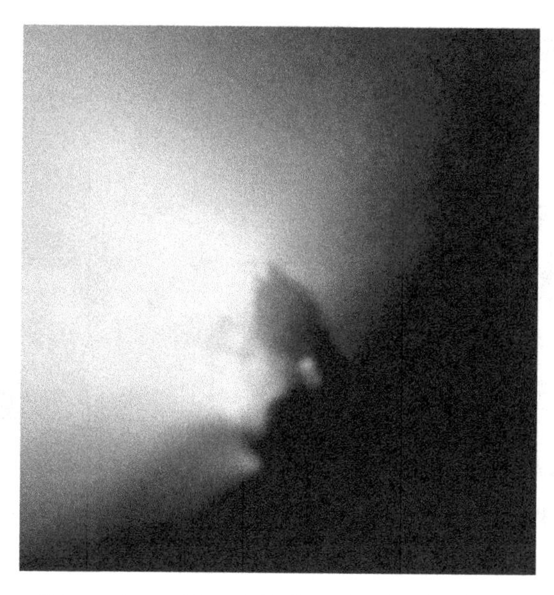

Figure 7.8 Image of the nucleus of Comet Halley taken by the Giotto spacecraft (courtesy: ESA).

The pressure of the gas streaming from the surface is enough to drag the smallest solid particles from the surface, but the larger ones are left behind, resulting in the gradual build-up of a layer of rubble. Since Comet Halley has now been around the Sun many times, there must now be very little ice in the surface layers and all the remaining ice must be hidden beneath this rubble layer. When the comet starts to approach the Sun again, the pressure under this layer will increase as the volatile material starts to sublime, and when the comet gets close enough to the Sun, jets of gas will break through weaker spots in this rubble crust.

Before we turn to the important question of where comets come from, it is impossible not to mention one final, very spectacular way that has been used to investigate the composition of a comet, which is to hit it with a hammer. In this case, the hammer was the 364 kg lump of copper dropped on Comet Tempel 1 (Figure 7.5) on 4 July 2005 as part of the Deep Impact space mission. The impact of the probe on the comet provided the biggest ever fireworks display for Independence Day (Figure 7.9), but the mission also had a serious scientific purpose. Observations of a comet's coma show only the material that has left the nucleus as the result of the heating effect of the Sun, and the NASA scientists hoped that the debris thrown up by the impact would allow them for the first time to look at *all* the material making up the upper layers of a comet. Observations of the cloud of debris by the Deep Impact spacecraft itself and by other telescopes showed that the upper 20 m of the comet contain a large amount of water ice, but that the metre immediately below the surface now contains very little ice. The observations also showed that

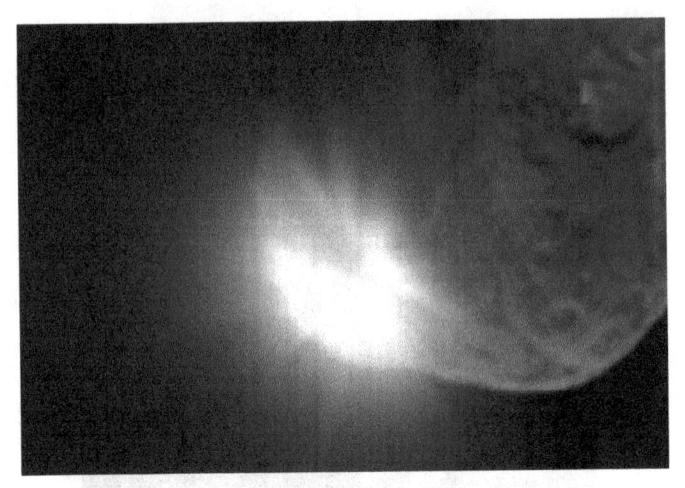

Figure 7.9 Image of the collision of the Deep Impact probe with Comet Tempel 1 taken from the Deep Impact mother ship (courtesy: NASA/JPL-Caltech/UMD).

this sub-surface ice does not consist of large slabs of ice like those seen on the Earth but rather of tiny grains about 1 μm in size. One intriguing result was that the spectroscopic observations revealed that the dust has a very different mineralogical composition from the dust brought from Comet Wild by Stardust. This suggests that there are large differences between individual comets, but it is also possible that the apparent difference is due to problems in the very different experimental methods used in the two missions.

7.4 The Oort Cloud

The properties of the comets we see today hold some important clues about where they come from. Comets fall into two main groups: *short-period comets* are ones with orbital periods less than about 200 years; *long-period comets* are ones with orbital periods greater than this value. The most well-known example of a short-period comet, of course, is Comet Halley, which returns to the inner Solar System every 76 years. Most short-period comets have orbital planes which lie fairly close to the ecliptic plane in which the planets orbit around the Sun, although their orbits are usually much more eccentric than those of the planets. A difference between short-period and long-period comets, which may indicate a different origin for the two groups, is that the orbits of long-period comets are not confined to the ecliptic plane – instead the long-period comets approach the inner Solar System from all directions. The long-period comets also generally have very long orbital periods, often greater than 1 million years. In 1950, the Dutch astronomer Jan Oort realized

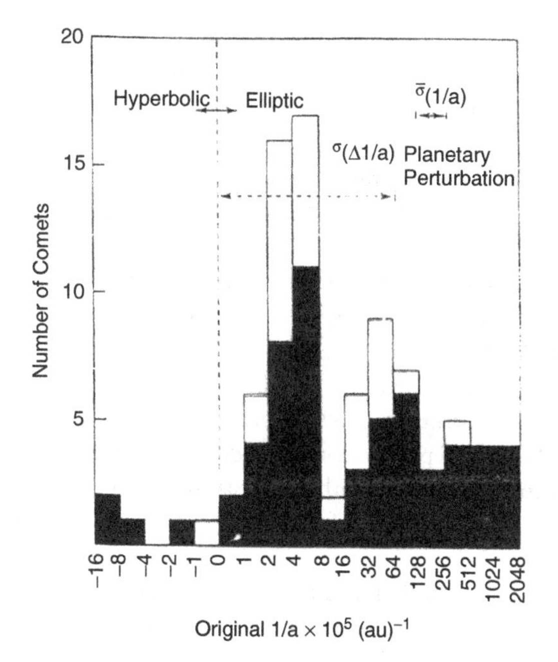

Figure 7.10 Histogram of the reciprocal of the semi-major axis of long-period comets (taken from Whipple, F.L. (1978) in *Cosmic Dust*, (ed. J. McDonnell), John Wiley & Sons).

that the detailed orbital characteristics of the long-period comets were clues to the origin of this class of comets, at least.

Figure 7.10 is a histogram of the reciprocal of the semi-major axis, a, of the orbits of the long-period comets. A positive value of $1/a$ indicates that the comet is on an elliptical orbit and is therefore bound to the Sun; a negative value shows that the comet is on an unbound hyperbolic orbit. Oort's first insight was he realized the lack of negative values (he attributed the few that do exist to measurement errors) shows the long-period comets are still members of the solar family and are not interstellar objects that pass through the inner Solar System by chance.

Oort's second insight came from the narrowness of the peak in the figure. When a long-period comet enters the inner Solar System, its orbit will be perturbed by the gravitational attraction of the giant planets, in particular Jupiter. Oort realized that if the long-period comets have repeatedly come through the inner Solar System, their orbits would have been so perturbed that we would not now see the narrow peak in the figure. The long-period comets must therefore be *new* comets, comets that are entering the Solar System for the first time and, once they leave, will never enter the inner Solar System again. He realized that if the current rate of new comets ($\approx$20 per year) has been the same since the formation of the Solar System, there must be somewhere a reservoir containing at least $20 \times 4.5 \times 10^9 \approx 10^{11}$ cometary nuclei.

Oort deduced that this reservoir must be spherical with inner and outer radii of about 25 000 and 200 000 AU. The reservoir must be spherical to explain the random directions of the long-period comets. Oort deduced the distance limits from the gravitational perturbing effect of nearby stars. If a cometary nucleus is beyond 200 000 AU from the Sun, the gravitational effects of other stars are likely to detach it completely from the Sun; if the cometary nucleus is between 25 000 and 200 000 AU from the Sun, the gravitational effect of nearby stars is large enough that there is an appreciable probability of it being moved onto an orbit that will take it into the inner Solar System, thus producing a new long-period comet; if the cometary nucleus is within 25 000 AU from the Sun, it will be so tightly bound to the Sun that nearby stars will have little effect on its orbit and it will never enter the inner Solar System (there may be cometary nuclei there but we would never see them). These radii, which Oort deduced purely from considering the gravitational effect of nearby stars, are also in good agreement with the measured semi-major axes of long-period comets in Figure 7.10. In honour of Oort, this spherical reservoir of comets is now called the *Oort Cloud*.

The Oort Cloud is a rather peculiar thing, because although the theoretical arguments that it exists are very strong, in the six decades since Oort's original paper nobody has thought of a way of observing it directly. Although the Oort Cloud must contain a very large number of objects, its total mass is probably not particularly large: approximately 10 times the mass of the Earth if an average cometary nucleus has a diameter of 10 km.

Another peculiarity is that the objects in the Oort Cloud, which may one day enter the inner Solar System, probably came from there in the first place. In the standard model of planet formation, which we will consider in the next chapter, it is very hard to see how the objects in the Oort Cloud could ever have formed where they are today. According to this model, the solid objects in the Solar System were formed from the gradual coalescence of the dust in the rotating disc of gas and dust that surrounded the primitive Sun. However, even if this disc extended all the way out to the Oort Cloud, its density there must have been so low that it would have taken an impossibly long time for the dust to coalesce. Fortunately, the standard model does provide a natural explanation for the existence of the Oort Cloud. Before the planets existed, the Sun was surrounded by *planetesimals*, chunks of rock and ice that had formed out of the disc but had not yet coalesced into planets. Many of these planetesimals eventually coalesced to form the planets, but even after the planets were formed there would still have been billions of planetesimals left over. Once the giant planets were formed, however, the orbits of most of these planetesimals would no longer have been stable, and computer simulations have shown that within 10 million years most of these planetesimals would have been ejected from the Solar System. The simulations suggest that most of the planetesimals escaped from the Sun completely, with about 10 % (the exact number is very uncertain) forming the Oort Cloud. The evidence that the Solar System was once filled with billions of planetesimals is still around us, because in the few places in the Solar System where the orbits of small objects are stable (the asteroid belt, Jupiter's Lagrangian points) we still see very large numbers of small objects.

7.5 The Edgeworth–Kuiper belt

The long-period comets therefore come from the Oort Cloud. For many years astronomers assumed that short-period comets were the descendants of long-period comets which had been captured on to short-period orbits by Jupiter's strong gravitational field. This plausible explanation was the one given in many textbooks, which just shows the danger of explanations which are plausible but which have not been carefully worked through. It began to unravel in the 1980s when it became possible to make detailed computer simulations of the paths of comets through the Solar System. The computer modellers discovered that too few short-period comets were produced by this process, and the short-period comets that were produced did not have orbits close to the ecliptic plane.

The true origin of the short-period comets only became clear in the next decade. On 30 August 1992, two astronomers from the University of Hawaii, Jane Luu and David Jewitt, were observing on the 88-inch telescope on Mauna Kea. They were carrying out a project to look for objects beyond the orbit of Neptune, using the simple technique of looking for objects that move relative to the fixed stars, the same one William Herschel had used to discover Uranus and Clyde Tombaugh had used to discover Pluto. Since Pluto, the only object then known to be beyond the orbit of Neptune, was discovered in 1930, this was very much a speculative long-shot project and Luu and Jewitt had already spent 5 years looking for these objects without any success.

On this night, however, after taking four exposures of the same piece of sky, they immediately realized they had discovered something interesting (Figure 7.11). Although most of the objects were in the same place in the four images, two objects had moved during the time between the exposures. One object was moving so quickly that it had moved even within the time of an individual exposure, producing streaks on the images. In the Solar System, an object close to the Sun moves more quickly than one further from the Sun, and Luu and Jewitt knew this object must be an asteroid, something in which they had little interest. The object that did interest them was the one that was moving slowly between the exposures. Using Kepler's third law (Exercise 1), they calculated that this object must be orbiting the Sun at a distance of approximately 40 AU, well beyond the orbit of Neptune, which has a radius of 30 AU. They were able to make a rough estimate of the diameter of the object from its brightness, using the method I described above. This was approximately 200 km.

This was the first *trans-Neptunian* (beyond the orbit of Neptune) object discovered in 60 years, but within a few months a second one had been discovered, and

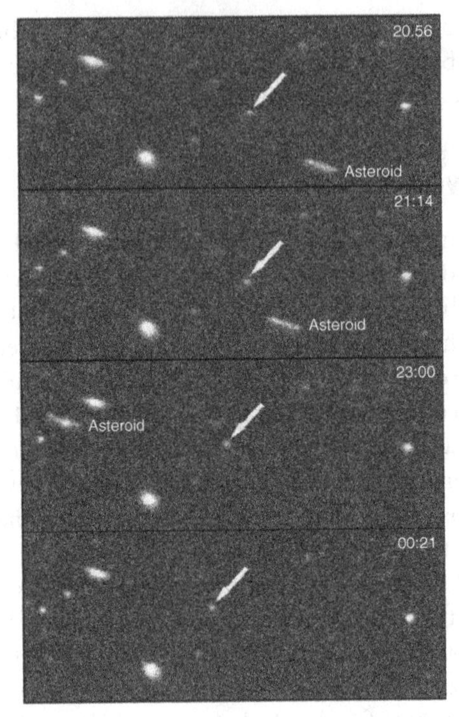

Figure 7.11 The four images taken by Jane Luu and David Jewitt that led to the discovery of the first trans-Neptunian object (courtesy: David Jewitt).

there are now over 1000 of these objects known. Luu and Jewitt had not discovered a planet, but a belt of objects like the asteroid belt in the inner Solar System. The existence of a belt of objects outside the orbit of Neptune was independently proposed by the Irish astronomer Kenneth Edgeworth in 1943 and the American astronomer Gerard Kuiper in 1951, and the belt is usually called either the *Kuiper belt* or, as I will do, the Edgeworth–Kuiper (EK) belt. Only a small part of the sky has been surveyed, and so the total number of objects in the belt is uncertain, but one recent estimate is that it contains approximately 100 000 objects with diameters greater than 100 km. The total mass in the belt is not large, one estimate being that the total mass of all the objects in the EK belt with diameters greater than 2 km is approximately 10 % of the mass of the Earth.

The orbits of the objects in the EK belt are generally close to the ecliptic plane, and so its discovery immediately suggested an alternative source for the short-period comets. Although the orbits of most of the EK objects are stable, it seems likely that occasionally an object will be dislodged from the belt, as the result of collisions and the gravitational effect of Neptune, and may thus move onto a trajectory taking it into the inner Solar System, becoming a short-period comet. This is currently the only credible explanation of the origin of the short-period comets, and additional evidence that it is correct is the existence of a group of objects that seem to be currently on their way from the EK belt to the inner Solar System.

The Centaurs are a group of approximately 25 small objects which have orbits between those of Jupiter and Neptune and which cross the orbits of one or more of the giant planets. Planet-crossing orbits like this are not stable and so these objects must be on a journey. The Centaurs have a large range of colours, including some very red ones, very similar to the range of colours seen in the EK belt. Given their colours and orbits, the only plausible point of origin is the EK belt. Computer simulations show that within 10 million years a Centaur is likely to pass so close to a giant planet that its orbit will be drastically changed, either taking it out of the Solar System completely or taking it closer to the Sun. Some evidence for the idea that the Centaurs may one day become short-period comets is that one object, 2060 Chiron, already shows evidence of cometary activity, exhibiting a faint coma. It therefore seems very likely that the Centaurs are objects that have been dislodged from the EK belt and, if they are not thrown out of the Solar System, will eventually become short-period comets.

So many objects have now been discovered in the EK belt that it is now clear there are at least three distinct groups of objects. Objects in the classical EK belt have approximately circular orbits with semi-major axes between 42 and 48 AU, far enough away from Neptune for the orbits to be stable. Resonant EK objects are ones that are closer to the Sun but have stable orbits because they are in an *orbital resonance* with Neptune. There are, for example, approximately 100 EK objects with an orbital semi-major axis of 39.4 AU, which means (from Kepler's third law) that they orbit the Sun twice in the time it takes Neptune to orbit the Sun three times – and so are in a 3 : 2 orbital resonance with Neptune. One of these objects is Pluto, which led to the naming of the objects in this class as *plutinos* and the gradual realization that Pluto is simply a rather large EK object (Chapter 1). Objects in the third class, the *scattered EK belt*, have highly eccentric orbits with small perihelia, typically about 35 AU, but aphelia of up to 200 AU. For most of their orbits these objects are too faint to be seen from the Earth, and so it is certain that there are many more objects in the class than we currently know about; one recent estimate is that the scattered and classical EK belts contain approximately the same numbers of objects.

The origin of the EK belt and the connections between the different classes are still unknown and hot topics for research. It is not clear, for example, whether the objects in the EK belt were formed where we see them today. As with the Oort Cloud, there is the problem that the density of the solar nebula at this distance from the Sun may have been too low for solid objects to coalesce. One possibility is that the density of the solar nebula was high enough for solid objects to form, but that we currently see much less material at this distance from the Sun because of the erosion of the EK objects by collisions during the last 4.5 billion years. The origin of the scattered belt is also unknown. Although the perihelia of many of its members lie close to the orbit of Neptune, suggesting that they are objects that strayed too close to the planet and were scattered outwards by the planet's gravitational field, there are a few objects in the scattered belt which have orbits that could not have been produced in this way. The giant EK object, Sedna, for example, which has a diameter approximately half that of Pluto, has an aphelion of almost 1000 AU and a perihelion of 76 AU, which is well beyond the gravitational influence of Neptune.

Rather surprisingly, we may already have detected, albeit indirectly, similar belts around other stars. In any belt like the asteroid or EK belt collisions will gradually erode the individual objects, and the endpoint of this continual pummelling will be tiny grains of dust. The Pioneer 10 and 11 spacecraft detected dust that is probably the debris of the collisions in the EK belt. The EK objects are cold and so emit much of their radiation in the infrared ($1 < \lambda < 100\,\mu m$) and submillimetre ($100\,\mu m < \lambda < 1\,mm$) wavebands, but even at these wavelengths they are still hard to detect from the Earth, and would be impossible to detect if they were around other stars. The amount of radiation, however, depends on the surface area of the source, and if an EK object is ground into dust it has much more surface area, so the dust will be a much stronger infrared and submillimetre source than the original object. Astronomers have used submillimetre telescopes to detect rings of dust around several stars (Figure 7.12). It thus seems likely that these belts are common and that this dust is the debris of the collisions in them.

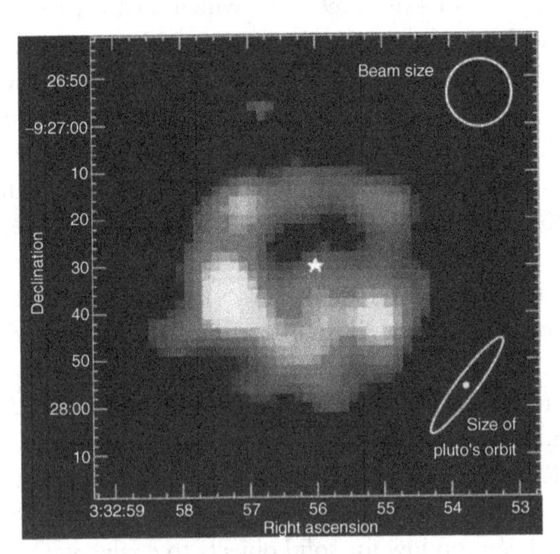

Figure 7.12 Submillimetre image of a ring of dust around the nearby star Epsilon Eridani. The ring peaks at about 60 AU from the star (courtesy: Jane Greaves).

Exercises

1 Suppose that you take two pictures of the same piece of sky 6 hours apart. Most of the objects on the two images are in the same positions, but one object has moved 3 arcsec between the two images. Estimate the object's distance from the Sun.

2 The radioactive decay of ^{87}Rb to ^{87}Sr has a half-life of 72 Gyr (1 Gyr is 10^9 years). Suppose that you have measured the amounts of the isotopes ^{87}Rb, ^{87}Sr and the non-radioactive isotope ^{86}Sr in two minerals within a meteorite. The values of the ratios $[^{87}Rb]/[^{86}Sr]$ and $[^{87}Sr]/[^{86}Sr]$ are 3.4 and 1.65 for one mineral and 4.4 and 1.695 for the second mineral. Calculate when the meteorite was formed.

3 (a) A comet has a nucleus with a diameter of 1 km and is at a distance of 1 AU from the Sun. On the assumption that most of the material lost from the comet is water vapour, estimate the mass of material lost from the comet each second. You may assume that its albedo is 0.05 (luminosity of Sun: 3.8×10^{26} W; latent heat of sublimation of ice: 2.8×10^6 J kg^{-1}).

(b) On the assumption that each time the comet returns to the inner Solar System it spends approximately 3 months at $\approx$ 1 AU from the Sun, make a rough estimate of how many times the comet will return before all the ice is melted (density of ice $\approx$1000 kg m^{-3}).

Further Reading and Web Sites

A'Hearn, M.F., Belton, M.J.S., Delamere, W.A. *et al.* (2005) Deep impact: excavating comet Tempel 1. *Science*, **310**, 258.

Brownlee, D., Tsou, P., Aléon, J. *et al.* (2006) Comet 81P/Wild 2 under a microscope. *Science*, **314**, 1711.

Luu, J. and Jewitt, D. (1996), The Kuiper Belt. *Scientific American*, May Issue.

Luu, J. and Jewitt, D. (2002) Kuiper belt objects: relics from the accretion disk of the sun. *Annual Reviews of Astronomy and Astrophysics*, **40**, 63.

http://www.ifa.hawaii.edu/faculty/jewitt/kb.html – the best web site about the outer solar system created by one of the discoverers of the Edgeworth–Kuiper Belt. Accessed 19 September 2008.

http://www.nasa.gov/mission_pages/deepimpact/main/index.html– the Deep Impact web site. Accessed 19 September 2008.

http://stardust.jpl.nasa.gov/home/index.html – the Stardust web site. Accessed 19 September 2008.

8 The origin of planetary systems

What we know is not much.
What we do not know is immense.
Pierre-Simon Laplace

8.1 Laplace's big idea

Any theory for the formation of planetary systems must explain some facts about our own planetary system. These facts are so easy to take for granted that it took a genius – in this case the Marquis de Laplace[1] – to realize that planetary systems could actually have been very different and that these facts need to be explained by any theory. Laplace decided there were four interesting properties of the Solar System that must be explained by any origin theory: (i) the orbits of all the planets are all roughly in the same plane; (ii) all the planets orbit around the Sun in the same direction; (iii) the orbits of all the planets are close to being circles; (iv) the planets spin on their axes in the same direction that they orbit around the Sun. Laplace was not quite right about the fourth fact because Venus and Uranus do not spin in the same direction as the other planets, but he was right that all these facts are significant – and, as I will show later in this chapter, even the anomalous rotation of two of the planets tells us something important about the origin of the Solar System.

[1] The philosopher Immanuel Kant had earlier proposed a similar theory but Laplace does not seem to have been aware of this. Kant was also a genius, of course.

Planets and Planetary Systems Stephen Eales

Laplace is a good example of the upwards mobility that a scientific career can bring, since he started life in humble surroundings before the French Revolution and ended up, after the restoration of the Bourbon monarchy, as the distinguished Marquis de Laplace. He is best known for his many contributions to mathematics. His theory for the origin of the Solar System, in contrast, contained embarrassingly little mathematics; he presented it almost guiltily as a footnote in his five-volume *Mecanique Celeste* 'with that uncertainty which attaches to everything which is not the result of observation and calculation'.

Laplace's great insight was that he realized these four basic facts could be explained if the Solar System formed out of a rotating cloud of gas. He realized that if a gas cloud collapses under the influence of gravity, any rotation will produce a centrifugal force that will cause the collapse to occur preferentially along the axis of rotation. As the cloud collapses, gravitational energy is converted into heat, which increases the pressure in the gas and eventually stops the collapse – with the end result being a rotating disc of hot gas. Laplace suggested that while the Sun formed in the centre of the disc, the planets were formed out of the surrounding material. He argued that as the disc of gas cooled, it would have separated into rings and the material in the rings would have gradually coalesced to form the planets. This simply theory naturally explained why the planets all move around the Sun in the same direction and why their orbits are almost circular and lie in the same plane. Laplace explained the fourth property as the result of the difference in speed between the gas at the inner and outer edge of each ring, which would have caused a planet to start to spin as it coalesced out of the ring material.

Despite his eminence, Laplace's idea was not universally accepted until late in the twentieth century. For one thing, as Laplace admitted, it was not a fully developed scientific theory. It also didn't provide an explanation of one other interesting fact about the Solar System: 99.9 % of its mass is in the Sun but 99 % of its angular momentum is in the planets. For most of the two centuries since Laplace's proposal there have been other competing theories. One alternative theory, for example, which was supported in the twentieth century by Sir James Jeans, was that a star had passed very close to the Sun and its tidal forces had drawn out a long filament of gas from the Sun, out of which the planets had formed. There are three reasons why Laplace's idea is now universally accepted.

First, if Laplace's idea, which is often called the *nebular hypothesis*, is correct, planets should be quite common, because a planetary system should form just about whenever a star is formed. In contrast, if the Solar System formed as the result of the chance encounter of the Sun with another star, planetary systems are likely to be quite rare. We now know, as the result of the discovery of the exoplanets (Chapter 2), that planetary systems are very common, and so the only tenable theories are ones in which planet formation is a fairly routine business. Second, we now know that the basic assumption of Laplace's theory – that a star forms as the result of the gravitational collapse of a cloud of gas – is correct, because radio astronomers have

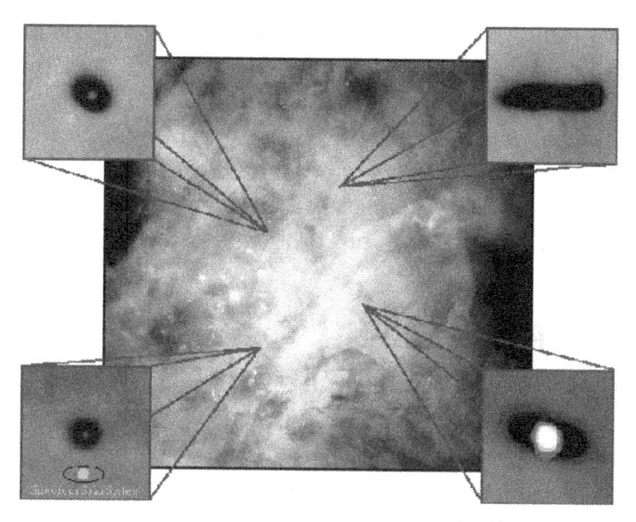

Figure 8.1 Image of part of the Orion Nebula taken with the Hubble Space Telescope. The insets show the silhouettes of dusty discs against the light of the nebula (reproduced courtesy of C.R. O'Dell/Rice University and NASA).

shown that the Galaxy is full of giant molecular clouds, and very young stars are always found in or close to these molecular clouds.

The most important piece of evidence for the nebular hypothesis, however, comes from observations of newly formed stars elsewhere in the Galaxy. Figure 8.1 shows an image taken with the Hubble Space Telescope of the Orion Nebula, the visible part of the Orion Molecular Cloud. By carefully studying this image, astronomers have discovered many dark patches, four of which are show as insets in the figure. In three of the insets there is a young star at the centre of the patch. These patches are the silhouettes of the discs around the stars, which are hiding the light from the background nebula because of the tiny solid particles (the dust) within the disc. In the top-right inset the disc is being seen edge-on and the dust is concealing the star itself. It should soon be possible to observe these discs directly because the dust emits far-infrared and submillimetre radiation; the Atacama Large Millimetre Array, which will start operation in 2010, will make it possible to study these discs in exquisite detail. The ubiquity of discs around young stars is one of the strongest pieces of evidence that the nebular hypothesis is correct.

Although the general idea that a planetary system forms out of a disc of gas and dust is now universally accepted, we are still ignorant of many of the details of how this happens, including ones which are rather more than details such as how the giant planets were formed. The basic problem is that the solar nebula was not a relatively simple system like a star, but a messy complex place, involving many different physical and chemical processes and all three phases of matter. We probably now have a fair understanding of most of these processes, although it is still possible we are missing some important ones, but we are still a long way from understanding how all these processes combined to produce a planetary system. In the rest of this chapter, I will describe the processes that are probably important in the formation of a planetary system. This is a rapidly advancing area of research, mainly because of the new information provided by observations of other planetary systems and of the discs around young stars. For example, as I will show below, the discovery of 'hot Jupiters' revealed the importance of one process that was missing from the models.

8.2 The protoplanetary disc

One thing we do know quite well is the birth date of the Solar System. The ages of most meteorites lie between 4.53 and 4.57 billion years (Chapter 7), and since these are fragments from asteroids and comets, objects that have been in deep freeze since the beginning of the Solar System, this must have been when the first large solid objects were formed. We will start our detailed discussion of the formation of the Solar System slightly before this time, when the collapse of the original gas cloud had stopped, but when no solid objects had yet formed out of the warm *protoplanetary disc*.

The collapse stopped because of the increase in pressure in the collapsing gas cloud due to the conversion of gravitational energy into heat. We can derive the structure of the disc using, once again, the principle of hydrostatic equilibrium. We will examine the balance between pressure and gravity for the small element of the disc shown in Figure 8.2. We will only consider forces in the vertical direction and we will assume the Sun has already formed in the centre of the disc and that all the gravitational force on this element is produced by the Sun, which is reasonable because the mass of the Sun is 99.9 % of the total mass of the Solar System.

Those without calculus should skip to Equation 8.4, which gives the relationship between pressure (P) and the height above the mid-plane of the disc (z) that must exist if the disc is not to collapse under its own weight.

The gravitational force on the element is

$$F_g = \frac{GM_{\odot}\rho A\delta z}{a^2} \tag{8.1}$$

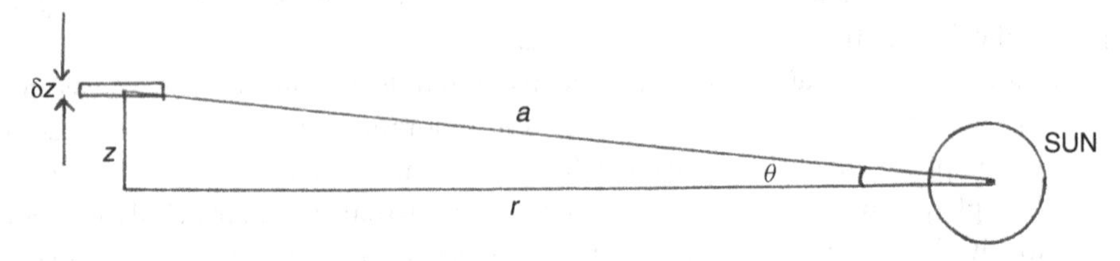

Figure 8.2 A small element in the protoplanetary nebula. r is the distance of the element along the disc, z is its distance in the vertical direction from the mid-plane of the disc.

in which A is the cross-sectional area of the element, δz is its thickness and ρ is the density of the gas it contains. We will assume the angle θ is very small, so $a \approx r$ and $\sin(\theta) \approx z/a \approx z/r$. The vertical component of the gravitational force is therefore

$$F_g = \frac{GM_\odot \rho A z \delta z}{r^3} \tag{8.2}$$

If the element is in equilibrium this force must be balanced by the force due to the pressure difference between the lower and upper face of the element: $A\delta P$. By equating the two forces and using the perfect gas law (Equation 5.2) to substitute for density, we can obtain the equation

$$\frac{\delta P}{P} = -\frac{\langle \mu_A \rangle m_{amu} GM_\odot z \delta z}{kTr^3} \tag{8.3}$$

in which T is the temperature, $\langle \mu_A \rangle$ is the mean molecular weight of the gas in the nebula and m_{amu} is the atomic mass unit. The negative sign arises because the pressure must decrease in the direction of increasing z for the gravitational and pressure forces to balance. By integrating this equation, we obtain the relation between pressure and height that must exist if the disc is not to collapse under its own weight:

$$P = P_0 e^{-\frac{z^2}{H^2}} \tag{8.4}$$

in which P_0 is the pressure at $z = 0$, the mid-plane of the disc, and H is given by

$$H = \sqrt{\frac{2r^3 kT}{\langle \mu_A \rangle m_{amu} GM_\odot}} \tag{8.5}$$

H is the scale height of the disc, the height at which the pressure has fallen by a factor of e^{-1}. The two equations show that the thickness of the disc increases with increasing distance from the Sun.

So far we have only been concerned with physics, but once the disc was in hydrostatic equilibrium it started to cool, and eventually material began to condense out of the disc – and chemistry became important. Figure 8.3 shows the result of one model of the sequence in which different elements condensed. The list along the top of the figure shows the 15 most abundant elements in the protoplanetary nebula, and directly below is a list of the compounds in which these elements were mostly found when the temperature of the nebula was 2000 K; most oxygen atoms, for example, were contained in water molecules. The staircase across the figure divides the solid phase from the gas phase, and by looking down the staircase from the top right to the bottom left we can follow the sequence in which the elements condensed as the nebula cooled.

The first elements that condensed were metals with high boiling points such as calcium, aluminium and nickel. Calcium and aluminium condensed as the oxides CaO and Al_2O_3 whereas nickel condensed as solid nickel. The arrows below the staircase show how these solids then reacted with other elements that were still in

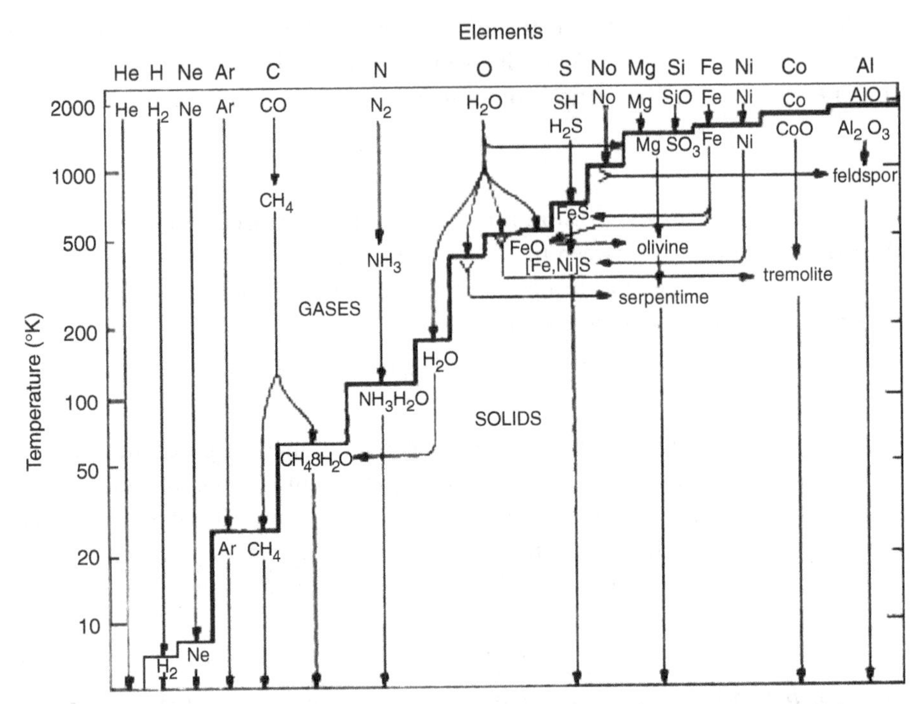

Figure 8.3 Predictions of the sequence in which chemical species condensed in the solar nebula. See the text for a detailed description of the diagram (reprinted, by permission, from the (1976) *Annual Review of Astronomy and Astrophysics*, **14**, 81, by *Annual Reviews*, www.annualreview.org).

the gas phase. According to this model, much of the aluminium, for example, ended up in the silicate mineral feldspar, and the presence in the diagram of silicates like feldspar and olivine is reassuring because most of the Earth is made of these minerals. Compounds with lower boiling points condensed as the nebula continued to cool, eventually even very volatile compounds such as water, ammonia and methane. The complexity of the entire process is shown by the possible routes for water molecules: water first started to enter the solid phase when the temperature was still relatively high by becoming incorporated in the hydrated minerals tremolite and serpentine, and only formed pure water ice when the nebula cooled further.

Despite the complexity of this model, evidence that something like this condensation sequence actually happened is two structures often seen within meteorites. *Chondrules* are small ($\approx$1 mm) round igneous rocks; their mineralogical properties imply they cooled very quickly – from a temperature of $\sim$1900 K to $\sim$1500 K within a few hours. *Calcium aluminium inclusions* (CAIs) are a little larger, typically a few millimetres in size, and light coloured (Figure 8.4). Both chondrules and CAIs must have formed when the temperature of the nebula was still very high, and it is strange

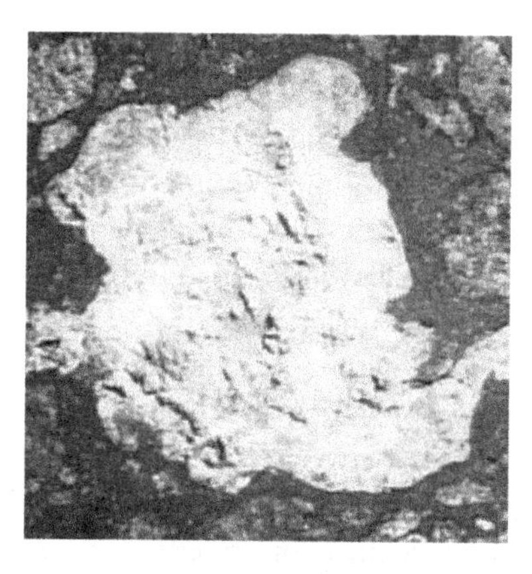

Figure 8.4 A CAI in the Allende meteorite.

to consider that these very tangible objects, which can be seen in most museum meteorite collections, date from a time before the planets even existed.

This model also partly explains one property of the Solar System of which Laplace was unaware: that there are two types of planet. The four planets closest to the Sun are essentially small rocky balls, the four outer ones giant balls of gas. The chemical composition of the two groups is different, with the inner planets composed mostly of the refractory elements that make up rock, such as silicon, magnesium and aluminium, and the outer planets largely made up of volatile elements. Of course, the difference in composition might be the consequence of the different sizes of the two groups of planets – large planets retain a larger proportion of volatile elements because of their stronger gravitational fields – but there is also a difference in composition between the moons in the inner and outer Solar System. Whereas our moon is made almost entirely of rock, the moons of Jupiter and of the other outer planets appear to contain a very large amount of ice (Chapter 4), suggesting there is a 'snowline' in the Solar System somewhere between the orbits of Mars and Jupiter. The condensation sequence in Figure 8.3 explains this difference rather nicely. Within the nebula there must have been a temperature gradient because of the heat from the Sun at its centre, and in the inner regions of the nebula the temperature would have been too high for volatile substances to condense, thus explaining why the inner planets are mostly made of rocky stuff and why Ganymede, for example, is covered by a layer of ice 700 km deep.

8.3 From dust to planetesimals

We now come to the most uncertain part in the formation of a planetary system. Once there were solid objects greater than about 1 km in size in the nebula, the physics of what happened next is fairly well understood. Once an object was this size, the gas in the nebula no longer had much effect on it and the only significant force was gravity. Because the physics is straightforward, theorists often start with a protoplanetary disc containing a large number of these kilometre-sized *planetesimals*, and then use computer simulations to investigate how gravity assembled these objects into planets. This, however, is the easy bit. The fundamental problem is how the planetesimals were formed in the first place – and here almost every aspect of the process is still uncertain.

The first uncertainty is in how the first solid particles stuck together. Computer simulations and observations of similar processes in laboratories on Earth suggest that the van der Waals forces between particles caused them gradually to stick together, forming fluffy clumps of loosely packed particles. This, however, is at odds with observations, because the chondrules seen in many meteorites must have formed very quickly and thus could not have been formed by particles gradually coalescing in this way.

The biggest uncertainties, however, come from the complex interactions between the solid particles and the gas in the nebula. There are two key processes, the second of which makes it surprising that there are any planets at all.

As I discussed earlier in this chapter, the reason the solar nebula did not collapse in the direction perpendicular to the disc was the gradient in the gas pressure. Solid particles, however, did not experience this pressure, and so the gas and the solid material began to separate: the gas remained in hydrostatic equilibrium while the solid particles sank under the influence of gravity towards the mid-plane of the disc. The speed at which a dust grain sank towards the mid-plane was limited by the viscosity of the gas and depended on the size of the grain – large grains sank faster. As a dust grain sank towards the mid-plane, the chance of it encountering other dust grains increased, and so it was likely to increase in size, which in turn made it fall faster. This was clearly quite a complex process. One simple calculation suggests that it took a grain with a diameter of 1 μm almost 10 million years to reach the mid-plane, which is longer than the total time that is generally assumed for the formation of the planets, but the dust grain's coalescence with other grains as it sank – a snowball rolling down a hill – makes it likely that the actual time was much less than this.

The second process operated in the direction of the Sun. The speed at which a planet orbits the Sun is found by balancing the centripetal force necessary to keep the planet moving in a circle with the gravitational force exerted by the Sun:

$$\frac{m_{\mathrm{p}}v^2}{r} = \frac{GM_{\odot}m_{\mathrm{p}}}{r^2} \tag{8.6}$$

and thus $v \propto r^{-1/2}$. This equation applied to solid objects after solid objects but it did not apply to the gas because it is likely there was a radial pressure gradient in the nebula partly balancing (for the gas) the inwards gravitational force. The effect of this pressure gradient was that solid objects moved faster than the gas. The velocity difference meant that any solid object effectively encountered a headwind as it moved around the Sun. Detailed calculations have shown that the effect of this headwind was greatest for objects of about 1 m in diameter: larger objects, such as planetesimals, did not even notice the headwind; smaller objects were more affected by the viscosity of the gas than the headwind. The effect of the headwind on a 1-meter-sized object would have been to cause it to spiral towards the Sun. The calculations imply that an object of this size initially orbiting the Sun at 1 AU would have fallen into the Sun in only $\approx$100 years, before it had a chance to coalesce with any other objects and making it hard to see how the planets were formed at all.

Planets do exist. The effectiveness of this process in removing metre-sized solid objects from the nebula shows that the formation of planetesimals must actually have occurred very quickly – otherwise they would not have formed at all. One possibility is that once a thin disc of solid particles had formed at the mid-plane of the nebula, it became gravitationally unstable and collapsed to form large numbers of planetesimals. Another possibility is that as the dust grains sank towards the mid-plane, the coalescence of dust grains into larger objects was extremely fast – snowballs rolling down a hill (except that in the inner part of the Solar System they were mostly balls of rock). The truth is nobody knows, and this is the most confused part of the story.

8.4 From planetesimals to planetary embryos

The story was much simpler once the objects were big enough that the effect of the gas was unimportant and gravity became the only important force. The gravitational forces between planetesimals led to frequent collisions, which had three possible outcomes: (i) the two planetesimals might bounce off each other; (ii) one or both might be broken into pieces; (iii) the two might coalesce. The outcome of a collision between a small and a much larger planetesimal was almost certainly coalescence, because even if the small one initially bounced off the larger one's surface it was unlikely to do so with a high enough velocity to escape from the larger one's gravitational field. Thus big planetesimals got bigger, and a simple argument shows that the biggest ones of all increased in size at the fastest rate.

The argument is very similar to the one used in the kinetic theory of gases to estimate the rate at which collisions occur between atoms. Let us suppose that the typical relative velocity between the solid objects in a protoplanetary nebula is v. If we ignore the effect of gravity, the mass of material swept up by a planetesimal each second is given by

$$\text{swept-up mass per second} = \rho_{\text{n}} \pi R^2 v \tag{8.7}$$

in which R is the radius of the planetesimal and ρ_{n} is the density of the solid material in the nebula (total mass of solid particles per unit volume of the nebula). This is equal to the rate of increase of the planetesimal's mass. Of course, we cannot completely ignore gravity, and a big planetesimal collides with more material because its gravitational field drags other objects towards it. In the 1960s, the Russian astrophysicist V.I. Safronov, who was responsible for much of the early work in this field, showed that the true rate of increase in the mass of a planetesimal is:

$$\text{Rate of mass increase} = \rho_n \pi v R^2 \left(1 + \frac{v_e^2}{v^2}\right) \tag{8.8}$$

in which v_e is the planetesimal's escape velocity. For small planetesimals, the second term in the brackets is unimportant and the rate of mass increase reduces to the one given in Equation 8.7. However, let us suppose that the planetesimal is large enough that the second term is larger than the first term, and so:

$$\text{Rate of mass increase} \simeq \rho_n \pi v R^2 \frac{v_e^2}{v^2} \tag{8.9}$$

We can obtain the escape velocity of the planetesimal by equating the gravitational potential energy at its surface with the kinetic energy of an object that just manages to escape from its gravitational field:

$$\frac{v_e^2}{2} = \frac{GM}{R} \tag{8.10}$$

Substituting this into Equation 8.9 and replacing the mass of the planetesimal by its volume times its average density, ρ_p, we obtain:

$$\text{Rate of mass increase} = \frac{8\pi^2 \rho_n \rho_p G R^4}{3v} \tag{8.11}$$

Thus while small planetesimals grow at a rate $\propto R^2$, large planetesimals grow at a much faster rate: $\propto R^4$. Because the planetesimal's mass depends only on the cube of its radius, the largest planetesimal in each region of the nebula rapidly outstrips all the others in that region, scooping up the lion's share of the solid material. In any region of the nebula there is therefore one planetesimal that becomes a *planetary embryo*.

8.5 From planetary embryos to planets

It is likely that the planetary embryos in the inner part of the nebula eventually ran out of solid material and stopped growing. We can estimate the maximum size of a planetary embryo using the idea of a *Hill sphere*, the region around an object in which its gravity is the dominant force. The radius of the Hill sphere, R_H, is defined as the distance from the object at which its gravitational force is just equal to the tidal force of the Sun (Equation 6.7). This is given by (Exercise 4):

$$R_H = \left(\frac{m_p}{3(m_p + M_\odot)}\right)^{\frac{1}{3}} a \tag{8.12}$$

in which a is the distance of the planetary embryo from the Sun, m_p is its mass and $M_\odot$ is the mass of the Sun. Detailed calculations suggest that a planetary embryo would have swept up all the solid material within about $4R_\mathrm{H}$, but this would still have left it well short of the mass of a planet. A planetary embryo at $\approx$1 AU from the Sun, for example, would have stopped growing when its mass was approximately six times the mass of the Moon, although this value depends critically on the density of the solid material in the nebula, which is very poorly known.

The rest of the story was a violent one. There were almost certainly many more planetary embryos than there are planets today. The orbits of the planetary embryos were continually changing because of the gravitational forces between them. Some were ejected from the Solar System completely; others merged to form the planets we see today. As I describe below, we can probably still see the results of one of these titanic collisions. It is difficult to carry out computer simulations of the entire process of planet formation, starting from the initial cloud of gas and finishing with planets, because there is still so much uncertainty about some parts of the process. It should be possible, though, to simulate the part that starts with planetesimals and finishes with planets, because the only force involved is gravity. Fortunately for the credibility of the story, simulations of this kind do produce results that look like real planetary systems.

This is the way the inner planets were probably formed. However, the outer planets do not obviously fit into this story because they are mostly balls of gas, and their origin is one of the biggest remaining questions. There are two alternative possibilities.

One possibility is that the cores of the outer planets formed in the same way as the inner planets. As the planetary embryo grew, it accreted gas as well as solid material from the protoplanetary nebula. The gas accretion rate depended on how fast the gas could cool once captured, which, according to detailed models, was not very fast when the embryo was small. The models show, however, that when the planetary embryo reached a mass equal to 10–20 times the mass of the Earth, the cooling rate rapidly increased, and the embryo would have suddenly accreted a very large amount of gas. In this *core-accretion* theory, the outer planets consist of a core of rock and ice, surrounded by an envelope of gas acquired when the core reached this critical mass.

The alternative theory, which was rejected for many years but which has now come into fashion again, is that the outer planets were formed by the sudden gravitational collapse of large parts of the protoplanetary nebula – very similar to the way the Sun itself was formed.

The evidence is inconclusive. If the core-accretion theory is true, the gas giants should contain solid cores, but as I explained in Chapter 4 the evidence for this is very weak. Another piece of evidence that has been suggested in support of this theory is that the abundance of heavy elements (ones with atomic weights greater than helium) is higher in the atmospheres of the giant planets than in the Solar

System as a whole, whereas if the gas giants formed by gravitational collapse their elemental abundances should be very similar to those of the solar nebula. However, this evidence, too, is inconclusive, because we only really know the abundances in the upper parts of the atmospheres of the giant planets (Chapter 5) and the upper layers of their atmospheres may have been contaminated by collisions with comets. A piece of circumstantial evidence in favour of the gravitational-collapse theory is that the gas giants, with their extensive systems of moon, do at least look like mini-Solar Systems.

8.6 Collisions, the Oort Cloud and planetary migration

Although there are still big gaps in the story, such as the origin of the gas giants, it does explain several other properties of the Solar System besides the four noticed by Laplace. V.I. Safronov showed in the 1960s that the different axes and rotational speeds of the planets could be explained by the collision of planetesimals with the growing planetary embryo, as long as the typical masses of the planetesimals were about 0.1 % of the mass of the final planet. If this is true, the length of day on Earth and the direction in space of the polar axis (the cause of the seasons) are the result of chance – the particular sequence of collisions that formed our planet 4.5 billion years ago.

As I discussed in Chapter 7, this scheme also provides at least a hand-waving explanation of the Oort Cloud and the two belts of small objects. After the formation of the planets, there were many leftover planetesimals. The orbits of most of these were unstable, and when sooner or later one of these planetesimals came too close to a planet, its orbit was modified by the planet's gravitational field so that it either crashed into a planet or was hurled beyond the orbit of Neptune. In this scheme, the objects in the Oort Cloud are the planetesimals that were thrown out of the inner Solar System and the objects in the asteroid and Edgeworth–Kuiper belts are the ones that did have stable orbits.

Collisions play a central role in this story, and the signs of impacts on most of the solid objects in the Solar System are additional evidence that this story is essentially correct. The most visible sign of the importance of impacts is, of course, the face of the Moon (Figure 8.5). A comparison of the dates deduced for different parts of the Moon's surface from 'crater counting' (Chapter 3) and from the radioactive dating of the rock brought back by Apollo shows that the rate of impacts was greatest during the first billion years of the Solar System's history, which is not surprising because this is when there would have been the greatest amount of builders' rubble left over from the construction of the planets. I described in Chapter 3 how the dark areas on the Moon's surface – the maria – can also be explained as the consequence of a few particularly violent collisions.

One of the remaining properties of the Solar System that needs to be explained is the existence of the Moon itself. Among the more than 100 moons in the Solar

Figure 8.5 Image of the Moon (courtesy: C.R. Lynds, KPNO/NOAO/NSF).

System, our moon stands out because the ratio of its mass to the mass of the Earth is so large (there are other large moons but they are around much bigger planets). There are also obvious explanations of the origin of all the other moons. The tiny moons of Mars are probably captured asteroids and some of the other moons with peculiar orbits, in particular Neptune's largest moon Triton, are also probably captured objects. As the giant planets with their extensive systems of moons *do* look like mini-Solar Systems, it seems likely that their moons were formed in the same process that formed the planets. The origin of our moon has been much more contentious.

Before Apollo, there were three theories for the origin of the Moon – all had major drawbacks. In the *capture theory*, the Moon was formed elsewhere in the protoplanetary nebula and was subsequently captured by the Earth's gravitational field. The problem with this theory was that the chance of the Earth capturing such a large object seemed very small. According to the *fission theory*, proposed by George Darwin, one of Charles Darwin's 10 children, the Earth was originally spinning so quickly that the centrifugal force caused it to split into two. The basic problem with this theory is that although we know the Earth is slowing down (Chapter 6), it seems very unlikely that it was ever spinning fast enough for this to happen. In the final theory, the *cocreation theory*, the Earth and the Moon somehow formed independently out of the same part of the nebula – exactly how they did this was unclear.

The Moon rock brought back by the various Apollo missions was additional evidence against the second and third of these theories. The detailed chemical analysis of the rock brought back by the astronauts revealed that it was different in one important way from rock on Earth: Moon rock contains a much lower proportion of volatile substances, substances with low boiling points. This difference was compatible with the first theory, but not with the other two, because if these are true the composition of the Earth and Moon should be very similar. However, this difference also inspired a new theory for the origin of the Moon.

According to this fourth theory, shortly after the formation of the Earth, a large object crashed into it. Computer simulations suggest this object must have been about the size of Mars. In this titanic collision a huge amount of material was gouged out of the Earth's mantle and thrown into space, and this debris eventually coalesced to form the Moon. This theory nicely explains why Moon rock contains few volatile substances because these would have boiled away in the heat produced by the impact. It also explains why the Moon has a lower density than the Earth, because the Earth has a dense core whereas the Moon was formed from lower density material from the Earth's mantle. In the decades since Apollo, this theory has gained almost universal acceptance among planetary scientists, because the importance of collisions is now so widely accepted and because computer simulations show that such a collision could have produced the Moon (Figure 8.6) – and also because there is no plausible alternative.

Other planetary systems must have formed in a similar way. We now have direct observational evidence that planetary systems are quite common (Chapter 2) and there is one nice argument based on the properties of the Solar System that there are probably large numbers of planets around every star. In the Solar System today, wherever there is the possibility of a stable orbit, we see either a planet or a group of smaller objects. Therefore the Solar System appears to be filled to capacity. This suggests that planet formation is so efficient that all planetary systems should be filled to capacity, and so when we are technically capable of detecting small rocky planets around other stars we should find them.

There is, of course, one important difference between the Solar System and other planetary systems. In the planetary systems that have so far been discovered, the giant planets are very close to their stars. It is still possible that most planetary systems are like our own, because these would be very difficult to detect using the Doppler method (Chapter 2), but there is no doubt we have to explain why there are any 'hot Jupiters' at all.

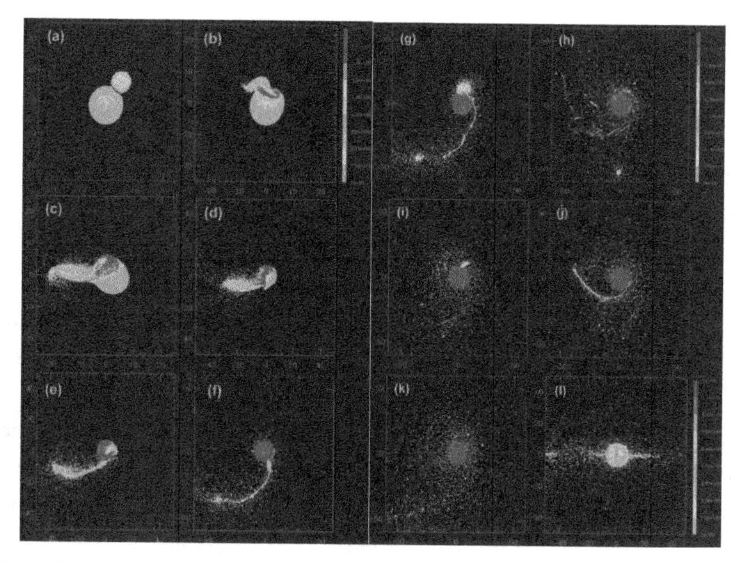

Figure 8.6 Computer simulation of a Mars-sized object hitting the Earth. The images show the Earth–Moon system at different times, from 6 minutes after the impact (a) to 27 hours (k). The final image (l) shows a side view of the system 27 hours after the impact. (reprinted, by permission, from the (2004) *Annual Review of Astronomy and Astrophysics*, **42**, 441, by *Annual Reviews*, www.annualreview.org).

The most popular explanation at the moment is *planetary migration*, which is a corollary of the most plausible explanation of the Oort Cloud – that it consists of planetesimals ejected from the Solar System. If a planetesimal is evicted from the inner Solar System as the result of a gravitational interaction with a planet, a requirement of the law of conservation of energy is that the planet itself must move to a lower position in the Sun's gravitational field. This change in position is very small, but after millions of encounters the planet may have moved a significant distance from its original position. The hot Jupiters, according to this theory, were formed much further away from their stars, but have since migrated inwards as the result of gravitational interactions with the leftover planetesimals. A possible explanation of why this does not appear to have occurred in the Solar System is if there were fewer planetesimals in the protoplanetary disc, perhaps because its mass was lower than those of the discs around other stars. Planetary migration also becomes complicated when more than one giant planet is involved; models of the Solar System suggest that while Jupiter did move some distance towards the Sun, Neptune actually moved outwards.

The story I have told in this chapter does seem to explain most of the properties of the Solar System, but I have tried to emphasize that it is not a fully developed scientific theory, but more a general scheme with parts that are still unfinished. I will finish the chapter with a reminder of one recent discovery that does not quite agree with the story. The material brought back by Stardust from Comet Wild (Chapter 7) contained crystals and minerals that formed at the high temperatures that existed in the inner parts of the nebula, whereas the ices in the comet must have been formed much further from the Sun. This discovery shows that even before the formation of planetesimals there must have been a significant flow of solid material in a radial direction – something that is not part of the story I have told here.

Exercises

1 A hydrogen atom falls in from infinity onto a protoplanetary nebula around a star with the same mass as the Sun. Suppose that the atom then moves in a circular orbit at 1 AU from the star. By comparing the change in gravitational potential energy with the kinetic energy the gas atom has from its circular velocity around the Sun, calculate the approximate temperature of the gas atom. (The average kinetic energy of an atom in a gas in thermal equilibrium is $(3/2)kT$.)

2 Suppose that rock consists of SiO, MgO and FeO. Let us suppose that in the solid matter that is formed in the protoplanetary nebula all three metals are in this form and any leftover oxygen that is not combined with the metals is contained in ice. Use the abundances in Table 1.2 to calculate the ratio of the mass of ice to the mass of rock in the Solar System (atomic weights of Si, Mg, Fe and O are 28, 24, 56 and 16).

3 (Calculus needed) Show that the radius, R, of a small planetesimal increases at a constant rate but the radius of a very large planetesimal increases at a rate proportional to R^2.

4 Derive an approximate formula for the radius of a planet's Hill sphere, the region in which its gravitational force is greater than the tidal force produced by the Sun. You should use Equation 6.7 to estimate the Sun's tidal force. The formula you derive will not be quite the same as Equation 8.12.

5 A planetary embryo in the inner Solar System reaches a maximum mass of about 6 times the mass of the Moon after it has consumed all the planetesimals within its reach. Estimate the approximate number of planetary embryos between the orbits of Venus and Mars. What are the possible fates of these objects? (Venus is 0.73 AU and Mars is 1.5 AU from the Sun; mass of Moon: 7×10^{22} kg.)

Further Reading

Gladman, B. (2005) The Kuiper Belt and the solar system's comet disc. *Science*, **307**, 71.

Soter, S. Are Planetary Systems Filled to Capacity? NASA on-line astrobiology magazine http://www.astrobio.net (accessed 19 September 2008; use the search facility on the web site to find the article).

Taylor, G.J. (1994) The scientific legacy of Apollo. *Scientific American*, **July,** 40–4.

9 Life in planetary systems

It is yet to be proved that intelligence has any survival value.
Arthur C. Clarke

9.1 A short history of life on Earth

The Copernican principle that we do not occupy a special place in the universe breaks down completely when we consider the origin of life. The argument that the Earth cannot be a special place implies that life should be a common phenomenon in planetary systems. But let us suppose that the probability of life arising is so small that there is only one planet in the Galaxy, perhaps only one in the universe, on which there is life. Since *we* are alive, we must be living on that lucky planet (winning the lottery does not seem improbable to the lottery winner). At the moment, therefore, until we discover a second example of life, there is no scientific evidence one way or the other; life might be incredibly common or we might be alone in the universe.

We are also hampered in any discussion of what life might be like on other planets because the only example we have is the life we know on Earth. The aliens in TV series like *Star Trek the Next Generation* always have two arms, two legs and two eyes and always look remarkably like handsome Hollywood actors, with just a touch of rubber and latex to make it clear they are honest-to-god extraterrestrials. It is easy to fall into this *anthropocentric fallacy* that life elsewhere in the universe should be just like life on Earth. But it is also possible that life elsewhere will have some of the same features that we see in life on Earth, because any forms of life living in a similar environment – for example, on a hard planetary surface in a gravitational field – will experience some of the same constraints and evolutionary pressures.

Planets and Planetary Systems Stephen Eales

We can make a little progress here by considering some circumstantial arguments. All life on Earth is based on the carbon atom, which can bond to four other atoms and can thus form many complex compounds, including those on which life is based. Some scientists have suggested that there might be a form of life based on the silicon atom, which can also bond to four other atoms. The details of the two atoms' electronic structure, however, mean that carbon chemistry is much richer than silicon chemistry, and of the 128 compounds that have so far been discovered in interstellar clouds[1] 96 contain carbon and only 8 contain silicon. It therefore seems likely that if life elsewhere consists of a self-replicating mixture of chemical compounds, and not on something more exotic such as the electric fields in ionized interstellar gas, it will be based on the same carbon chemistry we see on Earth. There is also an interesting argument based on evolution, because if we only have a single example of life, we do at least have the record of billions of years of life on Earth. For example, eyes appear to have evolved independently several times, and so because these appear to be a common evolutionary adaptation in a typical planetary environment, it seems likely that extraterrestrials (if they exist) will have eyes–so in this respect, at least, the producers of *Star Trek* may not have been hopelessly unimaginative. Nevertheless, despite arguments like this, my personal suspicion is that when we do encounter life for the first time beyond the bounds of our planet we are going to be totally surprised. In J.B.S. Haldane's words, 'the universe is not only queerer than we suppose, it is queerer than we can suppose'.

I should state now that I am not going to define what life is, because I could discuss this issue over several pages and nobody has ever been able to come up with a watertight definition anyway. I am simply going to assume that we will know life when we see it. In the rest of this chapter, I will first consider the history of life on Earth and only then discuss the possibility of there being life elsewhere in the universe. The history of life on Earth is important, not only because of the light it can shed on the possible existence of life elsewhere, but also because it shows there are many connections between the evolution of life and the evolution of the planet and planetary system in which it is found.

There is fossil evidence that life existed on Earth at least 3 billion years ago, and it may well have existed even earlier than this. Our knowledge of how life started on this planet, however, is still extremely poor and has not advanced much recently, which is shown by the fact that the classic experiment on this subject was carried out by Stanley Miller and Harold Urey over 50 years ago. In 1953 Miller, a new PhD student at the University of Chicago, and Urey, a professor at the university, designed an experiment to test the idea that the chemical building blocks of life might naturally have been formed in the early atmosphere of the primitive Earth, an idea originally proposed by the British geneticist J.D. Bernal and the Russian biochemist Alexander Oparin in the 1920s. The apparatus designed by Miller and

[1] http://www.cv.nrao.edu/~awootten/allmols.html (accessed 19 September 2008).

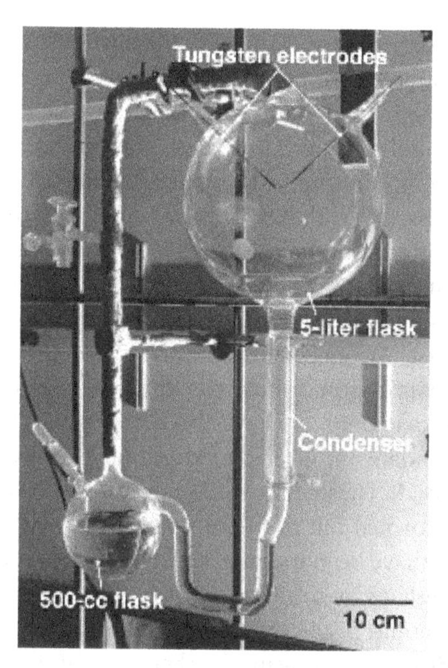

Figure 9.1 The apparatus used in 1953 by Stanley Miller and Harold Urey to test the idea that many of the chemical building blocks of life were naturally formed out of the compounds present in the original atmosphere of the Earth (see text for details).

Urey consisted of two glass flasks linked by two glass tubes (Figure 9.1). The upper flask contained the primitive Earth's atmosphere, which Miller and Urey assumed was a mixture of methane, ammonia, hydrogen and water, because they expected hydrogen-rich compounds to have been the most important ones in the atmosphere at this time. The water in the lower flask represented the Earth's oceans. By heating the lower flask to make water evaporate and cooling the bottom of the upper flask to make water vapour in this flask condense, they simulated the circulation of water vapour between the oceans and the atmosphere. They sent a continuous electrical discharge through the upper flask to simulate lightening bolts, which they thought might have been an important source of energy in the Earth's early atmosphere. After running the experiment for only a day, Miller noticed the water had turned pink. By the end of a week, it had turned a deep red. When he analysed the water he discovered that the colour was caused by a large number of organic compounds. The compounds that he found in this *primordial soup* included 13 of the 20 amino acids that are the building blocks for the proteins, some of the most important compounds in the human body. Later experiments of this kind have shown that other compounds important for life, including adenine, one of the four chemical bases that make up DNA, may also have been formed from the simple compounds that existed in the original atmosphere of the Earth.

Some planetary scientists have claimed that Miller and Urey's basic assumption about the primitive atmosphere of the Earth was incorrect, and that it was more likely to have been an oxidizing atmosphere of carbon dioxide, nitrogen and sulfur dioxide (Chapter 5). Similar experiments based on this kind of atmosphere do not synthesize organic compounds in anything like the quantities discovered by Miller and Urey. Nevertheless, even if they were wrong about the composition of the atmosphere, the importance of their experiment is that it showed that amino acids and other chemicals important for life are naturally formed out of simple common compounds. At least some of the building blocks of life seem to be present wherever we look in the universe. When scientists analysed the chemical composition of a meteorite that fell at Murchison in Australia in 1969, they discovered more than 70 different amino acids. Astronomers have also recently detected the spectral lines of guanine, the most abundant amino acid in Miller and Urey's chemical soup, towards an interstellar gas cloud. These discoveries suggest that even if amino acids and DNA bases were not formed out of the compounds present in the Earth's original atmosphere, they may have been manufactured elsewhere and then delivered to the Earth in some way, perhaps by comets.

One possibility I should briefly mention is that life itself was transported from elsewhere. The Swedish chemist Svante Arrhenius, who first proposed the idea in 1903, suggested that life was carried to the Earth by bacterial spores propelled across space by radiation pressure. Although the idea of *panspermia* seems to attract passionate supporters and opponents, with much mudslinging between the two camps, the truth is that there is no strong evidence one way or the other. It seems unlikely that bacterial spores would have survived the long voyage between the stars because of damaging radiation, but it is perfectly possible that life was transported to the Earth in some other way–in a comet, for example. Even if panspermia is correct, however, it does not answer the fundamental question of how life got started in the first place.

Amino acids and DNA bases, of course, are the building blocks of life, not life itself, and the biggest problem in explaining our existence is the step from the primordial soup to the first self-replicating organisms. Once these existed, the powerful engine of *natural selection* would have taken over, generating more complex organisms (although not very quickly, as we will see below).

One useful clue about the earliest forms of life comes from the genetic material of present-day organisms. Because genes mutate at a fairly constant rate, it is possible to determine the date when two species diverged: a small difference between the genes of the species (for example, a chimpanzee and *Homo sapiens*) implies that the common ancestor was recent; a large difference (for example, between a bird and a human) implies the two species diverged longer ago. By looking for common genes between the descendants, it is also possible to determine the genetic makeup of the common ancestor. Pushing this technique to the extreme, it should be possible, in principle, to recreate the genes of the ultimate ancestor: the ancestor of all the species on Earth. Although this technique has so far only been applied to a small part of the total genetic information in present-day organisms, the species today that appear to have genes most similar to those of this ultimate ancestor are microbes found in hot springs and in hydrothermal vents along mid-ocean ridges. This discovery suggests that life on Earth began in a hot environment.

Other than this, we are still stumbling around in the dark. The microbes in the hot springs and hydrothermal vents are bacteria, which, together with the Archaea, are the simplest independent organisms[2]. This simplicity is not too surprising if these were the first organisms to form out of the primordial soup. Both bacteria and archaea are *prokaryotes*, single-celled organisms that contain genetic material but do not contain a distinct nucleus, and so the creation of the nucleus must have been a later evolutionary step. One of the things that must have happened very early was the formation of cell walls. There are several ways this might have happened. One experiment has shown that soap-like compounds in the primordial soup might have formed into membrane structures rather like cells. It also seems possible that chemical reactions between compounds in the soup and solid surfaces were important, because these would have led to high chemical concentrations along the surfaces, and it would then only have been necessary for a membrane to form around one of these chemical concentrations to produce something that looked like a cell.

Another clue about the origin of life comes from a recent discovery about RNA, which is chemically slightly different from DNA and consists of a single-stranded molecule rather than the DNA double helix. In a cell the DNA and the proteins have very different roles. The DNA contains the genetic blueprint of the cell, while the main role of the proteins is to act as enzymes or catalysts for the cell's chemical reactions. As the DNA and proteins have completely different chemical structures, it seems very unlikely that the first cells contained both DNA and proteins, and so there is a chicken-and-egg problem: if the first cells contained only DNA, what catalysed the chemical reactions? And if they contained only proteins, what stored the information? The main role of the RNA in a cell is to transfer information from the DNA to the ribosomes, where the proteins are made. The big discovery about the RNA molecule is that not only does it store information, but it can also act as a catalyst for chemical reactions. A plausible solution to the chicken-and-egg problem is therefore that the first cells contained only RNA and that the primitive Earth was an 'RNA world'.

The rest of the history of life on Earth can be summarized in a paragraph. The first *eukaryotes*, unicellular organisms containing a nucleus, appeared between 1.6 and 2.1 billion years ago. At roughly the same time, organisms capable of extracting energy from sunlight by photosynthesis appeared. About 600 million years ago, the first complex multicellular organisms suddenly appear in the fossil record,

[2]Viruses are simpler but they are not independent because they require a host cell to grow and reproduce.

including many groups not seen today, in a spectacular efflorescence of life called the *Cambrian explosion*. Sixty-five million years ago roughly half the species on Earth were suddenly wiped out by the impact of an asteroid. Between 1 and 2 million years ago, the first recognizably human creatures were seen in Africa, with our species–*Homo sapiens*–appearing roughly 200 000 years ago.

There are several lessons we can learn from this very short history of the world. The first is that not much happened for an awfully long period of time. For roughly eight-ninths of the history of the Earth, the only creatures on our planet were single-celled organisms, which means that the first life we discover elsewhere is more likely to consist of bacteria than 'ET'. If intelligence is such a uniquely useful evolutionary adaptation, it has taken its time about appearing on this planet. It is always tempting to assume that our species is the inevitable culmination of evolution, but as Steven J. Gould has emphasized in his books[3] there is no 'progression' in evolution, merely an adaptation to the current environment. The suite of skills of *Homo sapiens*, including intelligence but also other qualities such as bipedal motion, clearly has some survival value at the moment, but we do not know how long it will continue to do so.

9.2 The evolution of the Solar System as a habitat

Another lesson is that the histories of life, the Earth and the Solar System are intertwined. The most obvious example is the effect of impacts in reorganizing life on Earth. There is now strong evidence that the disappearance in the fossil record 65 million years ago of about half the species on Earth, including the dinosaurs–the so-called 'KT extinction'–was caused by the impact of an asteroid. Everywhere on Earth where it is possible to see sedimentary rocks of the right age, there is a thin layer of clay about 1 cm in thickness. The clay is rich in iridium, an element that is rare on Earth but more common in meteorites. In the 1980s a team led by a father-and-son pair, the physicist Luis Alvarez and the geologist Walter Alvarez, claimed that one explanation of the extinction was if an asteroid had hit the Earth 65 million years ago. They argued that the debris thrown up by the collision would have obscured the Sun and shut down photosynthesis, and this and other calamities ensuing from the impact would have led to a large number of species becoming extinct–with the iridium-rich layer being formed when the debris settled back on to the surface. Using four different methods, all of which gave similar results, they estimated that the diameter of the asteroid was about 10 km. Their proposal was not immediately accepted for one obvious reason. The impact of an asteroid of this size should have made a crater about 300 km in diameter, and in the 1980s nobody knew of a crater of this size anywhere on the Earth. In the early 1990s, however, a crater of the right size and age was discovered under layers of sedimentary rock in the Yucatan peninsula of Mexico (Figure 9.2).

[3]For example, *The Burgess Shale and The Nature of History*.

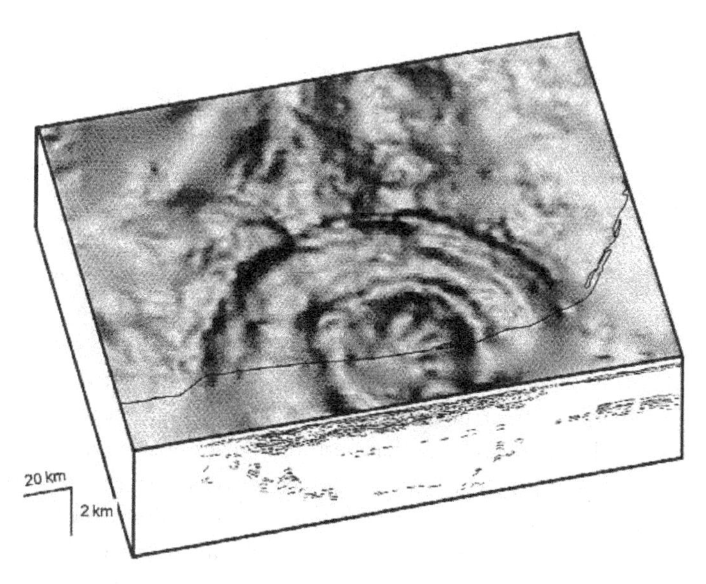

Figure 9.2 A perspective plot of the KT crater. The crater has a diameter of 180 km and is buried below a kilometre of sedimentary rock. This image has been created by mapping the small changes in the Earth's gravitational field on the surface above the crater (courtesy: Geological Survey of Canada).

The KT extinction is only the most recent extinction. In the fossil record there are extinctions roughly every 100 million years, which agrees quite well with the predicted interval between collisions with the Earth of asteroids of this size. The rate of collisions was much higher when the Earth was young, and some scientists have suggested that life could not have started until about 4 billion years ago because of the impact of objects large enough to vaporize the Earth's oceans. Impacts must be important in all planetary systems. The impact of an asteroid ten times larger than the one that produced the KT extinction would be enough to destroy all life on a planet, and it is possible that in some planetary systems life got started but was then destroyed by a planet-sterilizing impact.

There are several other ways in which life on Earth is linked to affairs beyond our planet. One of these is that the Earth's orbit is affected by the gravitational fields of the other planets. The result is that there are cyclical changes in the eccentricity of its orbit, in the inclination of its axis and in the direction this axis points in space. In the 1920s a Serbian civil engineer Milutin Milankovitch proposed that these changes, which cause the amount of sunlight hitting the Earth to vary, might explain the ice ages. Although the detailed agreement is not very good–during the last million years ice ages have occurred roughly every 100 000 years, which is not one of the Milankovitch cycles–it seems likely that at least part of the explanation is these orbital changes.

If subtle changes in the Earth's orbit have caused major changes to its climate, there is one change in the solar system that should have had a huge effect on life on Earth, but surprisingly does not seem to have done so. This is the fact that stellar models predict that the luminosity of the Sun has slowly increased during the last 4.5 billion years and that it is now 30% brighter than when it was formed. This change seems to have had very little effect on life on Earth.

We can estimate the size of this effect from Equation 1.5, which relates the temperature of the Earth to the luminosity of the Sun, which I give again here:

$$T = \left(\frac{L_{\odot}(1 - A)}{16\pi D^2 \varepsilon \sigma} \right)^{\frac{1}{4}} \tag{9.1}$$

As we saw in Chapter 1, the true temperature is slightly higher than this because of greenhouse gases, but if we assume that both the greenhouse effect and the albedo, which depends mostly on the cloud cover, have always been roughly the same, it is possible to use this equation to predict how the temperature of the Earth has changed as a result of the change in the Sun's luminosity. Figure 9.3 shows the prediction of a slightly more sophisticated model. This model predicts that until about 2 billion years ago the Earth's temperature was below the freezing point of water, and thus before this time the Earth should effectively have been a large snowball. But there is geological evidence that both water and life were present on the Earth at least 1 billion years before this.

One possible solution to the 'faint Sun problem' is if there is some feedback mechanism that has kept the temperature of the Earth relatively constant despite the gradual increase in the luminosity of the Sun. The most plausible possibility is the inorganic carbon cycle that I described in Chapter 3. On the Earth, most of the carbon dioxide, the most important greenhouse gas, is locked up in carbonate rocks as the result of the Urey weathering reaction, in which carbon dioxide dissolved in sea water reacts with calcium silicate in the rocks to form calcium carbonate. This is a two-way process, because carbon dioxide is returned to the atmosphere when a tectonic plate is forced under the surface; the rocks melt and the carbon dioxide is returned into the atmosphere by volcanic activity. The amount of carbon dioxide locked up in rock, and thus the strength of the greenhouse effect, depends on the balance between these two processes. Let us suppose that the Earth several billion years ago was very cold and the oceans were frozen. The Urey weathering reaction would have stopped. Volcanic activity, however, would have continued, and thus the amount of carbon dioxide in the atmosphere would have slowly increased, leading to an increased greenhouse effect and a higher temperature. This negative feedback loop seems a natural explanation of why the Earth's temperature was not much lower 3 billion years ago than it is today. If this argument is correct, it suggests that the existence of a system of tectonic plates is an important requirement if a planet is to be habitable, which in turn suggests that small planets such as Mars, which lose their internal heat too quickly, are unlikely to contain life. (The surprising constancy

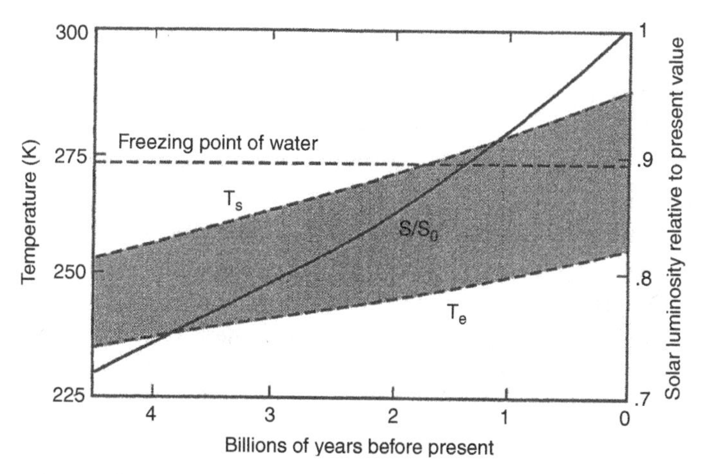

Figure 9.3 The predicted variation in the Sun's luminosity and the Earth's temperature during the last 4.5 billion years. The continuous line shows the gradual increase in the Sun's luminosity (see vertical axis on right). The lower dashed line shows the predicted temperature of the Earth without the effect of global warming, and the upper dashed line shows the predicted temperature if global warming is included in the model (reprinted, by permission, from the (2003) *Annual Review of Astronomy and Astrophysics*, **41**, 429 by *Annual Reviews*, www.annualreview.org).

of the Earth's temperature despite the gradual increase in the Sun's luminosity led James Lovelock, a British chemist, to suggest there must be feedback loops in the Earth similar to those in the human body. This, he argued, meant that the Earth itself is effectively a living creature, for which he coined the name Gaia. If Lovelock's controversial suggestion is right, there is a second type of life right under our feet.)

9.3 The possibility of life elsewhere

Now let us estimate the probability there is life elsewhere in the universe. I am going to assume this life, if it exists, is similar in one important way to life on Earth. One thing which gives some optimism for thinking there might be life elsewhere in the universe has been the discovery of organisms on Earth that are adapted to extreme environments. The so-called *extremophiles* have been found thousands of metres below the ice in Antarctica, in deep-sea hydrothermal vents at temperatures above the boiling point of water, and in extremely dry, acidic, alkaline and salty environments–and scientists have not yet found an environment on Earth that is so hostile that it is completely devoid of life. Organisms extract energy from the environment in a bewildering number of ways, from photosynthesis to the use of chemical energy in minerals like pyrites. There is, however, one respect in which all organisms on Earth are the same: all use water as a solvent. Even the extremophiles that exist at very high temperatures and low temperatures do this; the low-temperature cryophiles, for example, using protein 'antifreeze' to keep water liquid well below its normal freezing point. Although it has been suggested that life could be based on other solvents, such as ammonia, I will make the boring assumption that for there to be life on a planet it must be possible for water to exist in its liquid form.

One of the most famous equations in astronomy is one in which very few of the terms are known. In 1961 Frank Drake, a young astronomer at the National Radio Astronomy Observatory in Green Bank, West Virginia, wrote down an equation to estimate the number of intelligent civilizations in the Galaxy with which we might communicate. He realized this is the result of multiplying a large number of factors:

$$N = N_* f_p n_e f_l f_i f_c f_L \tag{9.2}$$

The first term in Drake's equation, N_*, is the number of stars in the Galaxy suitable for life. This is probably the best-known term in the equation, but even here there are uncertainties. There are about 300 billion stars in the Galaxy, but it seems unlikely, given the time it has taken for intelligent life to evolve on the Earth, that there could be life on planets around stars with very short lives, which rules out high-mass stars. It has also been suggested that life could not exist in the inner parts of the Galaxy because of the radiation from supernovae, which perhaps explains why we find ourselves living in the Galactic suburbs. I will assume, somewhat arbitrarily, that approximately one third of stars are in the *galactic habitable zone* and have long enough lives that they might harbour life, which means the first term in Drake's equation has a value of 100 billion.

The second term in the equation, f_p, is the fraction of stars that have planetary systems. We know from Chapter 2 that at least 5% of stars do have planetary systems, but the true factor might be much higher than this because the Doppler method is not sensitive to planetary systems like our own. I argued in the last chapter that the fullness of the Solar System implies planet formation is very efficient, and so I will assume that the value of this factor is one. In 1961, of course, Drake had no idea at all of the value of this term.

The third term in the equation, n_e, is the average number of planets in a planetary system that are suitable for life. This is the reason I made the assumption that liquid water is necessary for life because without it there is no obvious way to estimate this term. With this assumption, we can use Equation 1.5 to estimate the size of the region around a star in which liquid water could exist on the surface of a planet (Exercise 1). Although it is now apparent there are other places where liquid water might exist, such as the interior of Europa (Chapter 4), I will assume that life is only likely to start on a planetary surface. In the Solar System the Earth is clearly in this *habitable zone* but Mars and Venus are not (Table 1.1). Although there is no certainty that a planet would be found in the habitable zone around a star, the example of the Solar System suggests that a star's habitable zone is likely to contain at least one planet. I will therefore assume $n_e = 1$.

We now come to one of the most uncertain terms in the equation: f_l, the probability that life starts on a planet. Since it seems likely that life's chemical building blocks are found everywhere in the Galaxy, I will assume $f_l = 1$, but it is important to remember that as we still have very little idea of how these building blocks are put together, the true value of f_l might be minuscule.

The next term, f_i, is the probability that life, once it starts, eventually produces some intelligent species. Although it took 4.5 billion years for this to occur on the Earth, I will assume $f_i = 1$; intelligence clearly has some evolutionary value because otherwise we would not be here.

The next term, f_c, is the probability that the intelligent life form has both the means and the desire to communicate with us. Dolphins are intelligent but do not have the means, and some xenophobic extraterrestrials may not have the desire. I will ignore both possibilities and assume $f_c = 1$.

The last term, f_L, is the fraction of the star's lifetime during which an intelligent life form with both the means and the desire to communicate exists in the planetary system. In the Solar System, such a life form has existed for only about 60 years (since the development of suitable radio technology–see below), which is only a tiny fraction of the age of the Sun. Our estimate of the value of this term therefore depends entirely on how long we think a civilization like ours is likely to last. Drake wrote down his equation at the height of the Cold War, and at that time it probably seemed likely that the human race would soon sterilize the planet with nuclear weapons. Despite Arthur C. Clarke's cautionary words at the head of this chapter, the future now seems more cheerful. Given my inability to predict the future, I will try two alternative assumptions: a pessimistic one that human civilization, as the result of global warming or some other disaster, is destined to relapse into barbarism in about 200 years time; an optimistic one that the human race will solve its problems and retain its zest for scientific investigation until the Sun reaches the end of its life after about 5 billion years. In the optimistic case, the value of f_L is 0.5; in the pessimistic case it is 2.5×10^{-8}.

In the pessimistic scenario, I estimate that the number of civilizations in the Galaxy today with which we might communicate is 25 000; in the optimistic scenario, the number is 5×10^{10}, which would mean the Galaxy is currently teeming with life. However, the main point of a calculation like this is not to come up with an accurate estimate, but to reveal our areas of ignorance, and although we know a little more than Drake did in the early 1960s, there are still no convincing ways of estimating the fourth and final terms in the equation. Are there any other ways we can try to answer one of the biggest of all human questions: are we alone in the universe?

The famous Italian physicist Enrico Fermi put forward an interesting argument about the existence of extraterrestrial life. Inspired by the speed with which North America was settled in the nineteenth century, Fermi argued that once interstellar space flight became possible the Galaxy would fill up very fast. He claimed that some possible methods of interstellar space flight, such as large slow-moving 'arks' that take decades to move between the stars, do not seem that far beyond the technological horizon; and if there are many technological civilizations in the Galaxy, at least one of them should already have acquired this technology. But if so, where are they? Even with the slowest kind of interstellar space flight, it should take no more than a few million years to fill up the entire Galaxy, and so representatives of this civilization should already have visited the Solar System. Since there are no records that this has ever happened, Fermi claimed that it is likely we are currently the only technological civilization in the Galaxy.

There are many possible objections to *Fermi's paradox*. Possibly such an advanced civilization would not be driven by the same urges that drove the settlement of the Americas. Another possibility is that the Solar System has been visited but for altruistic reasons the extraterrestrials have been careful that we should not be aware of this (I personally do not believe any of the reports of extraterrestrial sightings on the grounds that if the extraterrestrials wanted to talk to us they would land their spaceship outside the United Nations building rather than try to talk to some redneck on a back-country road.)

At present, there is only one way that scientists have come up with for trying to answer this question, which is to look for radio signals from these civilizations. Frank Drake himself started this game when he used one of the radio telescopes at the National Radio Astronomy Observatory to look for signals from two nearby stars, Tau Ceti and Epsilon Eridani. Drake spent two months observing the stars but he did not detect any radio signals that could be messages from another civilization.

There are three main problems with radio SETI (*S*earch for *E*xtra*t*errestrial *I*ntelligence) programmes like this. The first is the huge range of possible frequencies. Drake made the decision to observe at a frequency of 1.4 GHz, close to the frequency of the spectral line emitted by atomic hydrogen, on the grounds that an extraterrestrial civilization might choose this as a natural communications frequency. The second problem is that even if there were 100 civilizations currently in the Galaxy, one would have to monitor about 1 billion stars to have a reasonable chance of detecting a single one. But the biggest problem of all is the third one. These radio searches rely on there not only being a civilization around a star, but on that civilization choosing to transmit a radio signal towards the Sun, 1 of 300 billion stars in the Galaxy, which does seem a trifle improbable. This is sometimes called the 'what if everyone is listening and nobody is talking' problem.

Since Drake's observations in the early 1960s, radio searches have improved in sophistication and scope. With new radio telescopes such as the Allen Telescope Array (Figure 9.4), which will be run by the privately funded SETI Institute, it will be possible to monitor 1 million stars on a billion different frequencies, thus going a long way to overcoming the first two problems. In the long term, it may even be possible to overcome the third problem. Since the beginning of the radio age,

Figure 9.4 An artist's impression of the Allen Telescope Array. When it is completed, the telescope will consist of 350 dishes and will be the first telescope whose main purpose is to look for radio signals from extraterrestrial life (reproduced courtesy of SETI Institute).

we have been inadvertently broadcasting to the universe (including some rather embarrassing material). Other civilizations may be broadcasting in a similar way, and the advantage of these signals is that they are broadcast in every direction; we do not have to rely on the civilization choosing to transmit a signal towards us. These signals are too faint to detect with current radio telescopes, but when the next-generation radio telescope, the Square Kilometre Array, is completed in about 2015 it will be possible to detect these signals from at least the closer stars. If we detect a signal, it will be one of the biggest events in human history.

Exercises

1 Estimate the inner and outer radius of the habitable zone for a star with a luminosity that is 100 times the luminosity of the Sun.

Further Reading and Web Sites

Alvarez, L.W., Alvarez, W., Asaro, F. and Michel, H.V. (1980) Extraterrestrial Cause for the Cretaceous-Tertiary Extinction. *Science*, 208, 1095.

Alvarez, Walter *T. Rex and the Crater of Doom*, Penguin Books, 1989.

Drake, F. A Reminiscence of Project Ozma. http://bigear.org/CSMO/HTML/CSframes.htm (accessed 19 September 2008).

http://www.seti.org/–web site of SETI institute (accessed 19 September 2008).

Answers

The full workings of the answers to all the exercises can be found at http://www.astro.cf.ac.uk/pub/Steve.Eales/index.html

1.1 Temperature ≈ 1 K. (This is not actually a realistic answer because the lowest possible temperature in the universe today is 2.7 K, the temperature of the cosmic background radiation, but it just goes to show that at the distance of the Oort Cloud the heating effect of the Sun is rather small.)

1.2 Thickness of Martian lithosphere $= 93$ km. This is higher than the value one obtains from the same calculation for the Earth, which suggests the reason plate tectonics does not occur on Mars is that the lithosphere of Mars is thicker than the Earth's.

1.3 The radius of the cavity is 400 km. The pressure at this radius is $\approx 10^9$ N m^{-2}, well below the tensile strength of iron.

2.1 Magnitude change when the planet passes in front of the star is 9.08×10^{-5}.

2.2 Velocity $= 0.09$ m s^{-1}.

2.3 The planet is $\approx 2.3 \times 10^{-10}$ times fainter than the star.

3.1 Time ≈ 100 million years. Parts of the continental crust are much older than this because when a continental plate and an oceanic plate move towards each other, it is the heavier oceanic plate that is forced down into the asthenosphere.

3.2 Thickness of continental plate ≈ 39 km.

4.2 Density of core ≈ 9000 kg m^{-3}.

4.3 The difference in temperature is ≈ 15.6 K.

4.4 Original temperature of Earth $\approx$1063 K.

5.1 Height at which pressure is 20 % of that at the surface $\approx$45 km. The difference in the temperature of the two objects means that the percentage of molecules that are travelling above the escape velocity is much lower on Titan than Mars.

5.2 Height of cloud layer $\approx$1700 m.

5.3 Mass of atmosphere $\approx 3.6 \times 10^{21}$ kg; thickness of rocks $\approx$9 km.

5.4 Radius of storm system $\approx$250 km.

6.1 Maximum radius of asteroid $\approx$32 km.

6.2 The angular diameter of the Moon will have decreased by $\approx$ 16 %.

6.3 (a) The tidal force per unit mass $\approx 6.6 \times 10^{-4}$ N; (b) the total force acting across the cross-section of the comet $\approx 1.6 \times 10^{15}$ N; (c) the total tidal force $\approx 8.2 \times 10^{9}$ N. The total tidal force is much less than the internal force holding the comet together, and so if the real comet had been made up of solid rock, Jupiter's tidal forces would not have been enough to disrupt it. Its real internal structure must therefore have been very different.

7.1 The object is approximately 44 AU from the Sun.

7.2 The meteorite was formed 4.56 Gyr ago.

7.3 (a) The amount of ice lost every second $\approx$ 358 kg; (b) the number of returns before all the ice is lost $\approx$ 200.

8.1 The temperature of the gas $\approx$ 36 000 K.

8.2 Ratio of mass of ice to mass of rock $\approx$1.4.

8.3 Use the chain rule to show that $\mathrm{d}M/\mathrm{d}t \propto R^2\mathrm{d}R/\mathrm{d}t$ and then use Equations 8.7 and 8.11.

8.4 Using this simple derivation, the radius of the Hill sphere is given by

$$R_{\mathrm{H}} \simeq \left(\frac{M_{\mathrm{p}}}{4M_{\odot}}\right)^{\frac{1}{3}} a$$

8.5 The number of planetary embryos $\approx$20. Some of these will be incorporated in the planets, others will be ejected from the Solar System.

9.1 Inner radius $\approx$2.2 AU and outer radius $\approx$4.1 AU (I have assumed an albedo of 0.5.)

Appendix A

A.1 The epoch of planetary exploration

The list below includes only the most important missions, at least as I see them. The date is the one on which the spacecraft reached the planet or moon, rather than the date on which it was launched, which can make a huge difference, especially for missions to the outer Solar System. I have only included future missions if they have already been successfully launched and are now on their way.

Mission	Importance
1968 (Apollo 8, USA)	First human voyage to another celestial body
1969 (Apollo 11, USA)	First human landing on another celestial body
1969 (Venera 7, Russian)	Mission to Venus – first successful landing on another planet
1971 (Mariner 9, USA)	First detailed images of Mars, which reveal Valles Marineris canyon system, huge volcanoes and channels cut by water
1974 (Mariner 10, USA)	Mission to Mercury, which produces images of 45 % of the planet's surface, revealing a heavily cratered surface like the Moon's
1976 (Viking 1 and 2, USA)	Mars mission that carries the first experiments to look for life on another planet (with ambiguous results)

Mission	Importance
1973–1989 (Pioneer 10 and 11, Voyager 1 and 2, USA)	First missions to Jupiter and Saturn – first detailed images of the planets and their moons; discovery that Jupiter has a ring system
1986 (Voyager 2, USA)	First spacecraft to visit Uranus, producing the first images of the planet (which looks like a star from the Earth). The planet looks quite different from Jupiter and Saturn, being blue and rather featureless. Ten new moons are discovered
1986 (Giotto, European Space Agency)	Mission to Comet Halley – first images of the nucleus of a comet
1989 (Voyager 2, USA)	First spacecraft to visit Neptune, producing the first images of the planet (Neptune looks like a star from the Earth). The planet is blue like Uranus but with a large dark spot. Six new moons and a ring system are discovered
1990 (Magellan, USA)	Mission to Venus, which uses radar to map the surface of the planet
1995 (Galileo, USA)	Mission to Jupiter, which makes detailed observations of the moons, discovering Ganymede's magnetic field and launching a probe into Jupiter's atmosphere
2004 (Cassini, USA and European Space Agency)	Mission to Saturn. The spacecraft discovers lakes on Titan (Chapter 5). The Huygens probe lands on the surface of the moon, the first landing on the moon of another planet
1999 (Mars Global Surveyor, USA)	Mission to Mars, which produces detailed images of the surface, a topographic map, and observations of the surface minerals (Chapter 3)
2003 (Mars Express, European Space Agency)	Mission to Mars, which is producing detailed images of the surface, mapping the distribution of important minerals (Chapter 3) and using radar to probe below the surface (Chapter 3)

Mission	Importance
2004 (Mars Exploration Rovers, USA)	Robotic geologists, which continue to study the detailed geology of two small regions of Mars
2004 (Stardust, NASA)	Mission to Comet Wild 2 that collected material from the coma and brought it back to Earth
2006 (Deep Impact, NASA)	Mission to Comet Tempel 1 that dropped a large weight on the comet. Observations of the debris revealed much about the interior of the comet (Chapter 7)
2006 (Venus Express, European Space Agency)	Mission to Venus using many of the same instruments as Mars Express
2006 (Mars Reconnaissance Orbiter, NASA)	Mission to Mars containing high-resolution cameras for observing the surface, radar for observing under the surface and spectrometers for mapping the minerals on the surface
2008 (Phoenix, USA)	Mission to Mars to study the soil in the northern arctic regions, in particular to measure the water content of the soil and to look for organic compounds
2011 (Messenger, USA)	Mission to Mercury
2011–2015 (Dawn, NASA)	Mission to the asteroid belt, which will arrive at Vesta in 2011 and at Ceres in 2015
2014 (Rosetta, European Space Agency)	Mission to Comet 67P/Churyumov-Gerasimenko, which is currently in the outer Solar System. Rosetta will release a small probe which will land on the comet. The mother ship will stay close to the comet and study it as it approaches the Sun
2015 (New Horizons, NASA)	Mission to the Pluto–Charon system

Appendix B

B.1 Derivation of Kepler's first and second laws

The natural coordinate system for considering the motion of a planet around the Sun is a polar coordinate system: r is the distance from the planet to the Sun and θ is the angle between the line joining the planet and the Sun and a reference direction (Figure B.1).

The only force on the planet is gravity, which means there is acceleration only in the radial direction. This has two components: the centripetal acceleration due to the planet's motion around the Sun, $r(\mathrm{d}\theta/\mathrm{d}t)^2$, and the acceleration due to the change in the radial coordinate, $\mathrm{d}^2r/\mathrm{d}t^2$. From Newton's second law ($F = ma$) and the law of gravitation, we obtain

$$\frac{\mathrm{d}^2 r}{\mathrm{d}t^2} - r\frac{\mathrm{d}\theta}{\mathrm{d}t}^2 = -\frac{GM_\mathrm{s}}{r^2} \tag{B.1}$$

The planet must also obey the law of conservation of angular momentum:

$$\frac{\mathrm{d}}{\mathrm{d}t}\left(r^2\frac{\mathrm{d}\theta}{\mathrm{d}t}\right) = 0 \rightarrow r^2\frac{\mathrm{d}\theta}{\mathrm{d}t} = h \tag{B.2}$$

in which h is the angular momentum per unit mass. Kepler's second law can be deduced from these equations quite simply; Kepler's first law can be derived with more effort.

Suppose that the planet moves a small angular distance $\delta\theta$ in a small time δt. The area swept out by the line joining the planet and the Sun (the hatched area in Figure B.1) is from simple geometry $(r^2/2)\delta\theta$. The rate at which area is swept out by this line is thus $(r^2/2)\delta\theta/\delta t$, which from Equation B.2 we can see is a constant, which proves Kepler's second law.

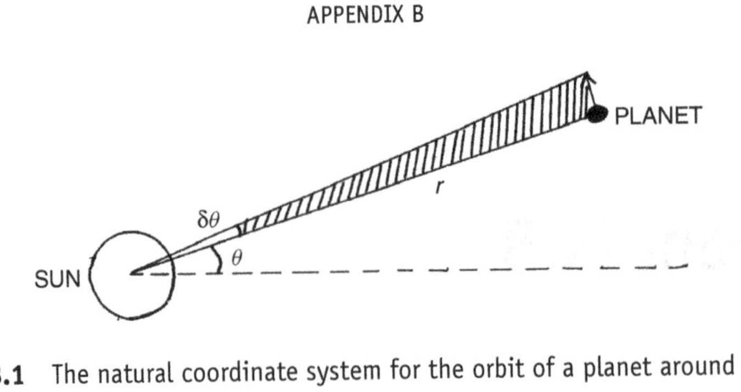

Figure B.1 The natural coordinate system for the orbit of a planet around the Sun.

To derive Kepler's first law, we start with the substitution $r = 1/u$. Using the chain rule, we obtain:

$$\frac{dr}{dt} = -\frac{1}{u^2}\frac{du}{dt} \quad \text{and} \quad \frac{d^2r}{dt^2} = -\frac{1}{u^2}\frac{d^2u}{dt^2} + \frac{2}{u^3}\left(\frac{du}{dt}\right)^2 \tag{B.3}$$

Using these substitutions, we can rewrite Equations B.1 and B.2 as

$$-\frac{1}{u^2}\frac{d^2u}{dt^2} + \frac{2}{u^3}\left(\frac{du}{dt}\right)^2 - \frac{1}{u}\left(\frac{d\theta}{dt}\right)^2 = -GM_s u^2 \tag{B.4}$$

$$\frac{d\theta}{dt} = u^2 h \tag{B.5}$$

The next step is to convert the derivatives of u with respect to time into derivatives with respect to θ using the chain rule:

$$\frac{du}{dt} = \frac{du}{d\theta}\frac{d\theta}{dt} = \frac{du}{d\theta}u^2 h \tag{B.6}$$

The right-hand equation follows from Equation B.5. We can convert the second derivative of u in a similar way:

$$\frac{d^2u}{dt^2} = \frac{d\theta}{dt}\frac{d}{d\theta}\left(\frac{du}{d\theta}u^2 h\right) = u^4h^2\frac{d^2u}{d\theta^2} + 2u^3h^2\left(\frac{du}{d\theta}\right)^2 \tag{B.7}$$

We can now use Equations B.5, B.6 and B.7 to rewrite Equation B.4 in a rather simple form:

$$\frac{d^2u}{d\theta^2} + u = \frac{GM_s}{h^2} \tag{B.8}$$

The general solution of Equation B.8 is

$$u = \frac{GM_s}{h^2} + A\cos(\theta - \theta_0) \tag{B.9}$$

in which A and θ_0 are the constants of integration. You can check that this is the solution by going backwards: differentiate B.9 twice and see if you get back to

Equation B.8. We now reintroduce r by reversing the substitution: $u = 1/r$. With a little rearrangement, Equation B.9 becomes

$$r = \frac{\dfrac{h^2}{GM_s}}{1 + \dfrac{Ah^2}{GM_s}\cos(\theta - \theta_0)} \tag{B.10}$$

This is identical to the equation of a conic section (an ellipse, parabola or hyperbola):

$$r = \frac{p}{1 + e\ \cos(\theta - \theta_0)} \tag{B.11}$$

with

$$p = \frac{h^2}{GM_s} \quad \text{and} \quad \mathrm{e} = \frac{Ah^2}{GM_s} \tag{B.12}$$

The path of any object in the Sun's gravitational field is therefore an ellipse, a parabola or a hyperbola. If $0 \leq \mathrm{e} < 1$ the object has an elliptical orbit (e then becomes the eccentricity of the orbit); if $\mathrm{e} > 1$ the object is not bound to the Sun and has a hyperbolic trajectory; if $\mathrm{e} = 1$ the object has a parabolic trajectory.

Index